India's Scientific Heritage
Founder Editor : Late Dr L M Singhvi

13

VEDIC PHYSICS

KESHAV DEV VERMA

VEDIC PHYSICS

Towards Unification of Quantum Mechanics and General Relativity

Foreword
by
DR MURLI MANOHAR JOSHI

Founder Editor's Exordium
by
L M SINGHVI

Introduction
by
DR S N BHAVSAR

Script assented
by
PROF P S RATHOR

MOTILAL BANARSIDASS PUBLISHERS
PRIVATE LIMITED • DELHI

First Edition: Delhi, ***2008***

ISBN: 978-81-208-3270-1

MOTILAL BANARSIDASS
41 U.A. Bungalow Road, Jawahar Nagar, Delhi 110 007
8 Mahalaxmi Chamber, 22 Bhulabhai Desai Road, Mumbai 400 026
236, 9th Main III Block, Jayanagar, Bangalore 560 011
203 Royapettah High Road, Mylapore, Chennai 600 004
Sanas Plaza, 1302 Baji Rao Road, Pune 411 002
8 Camac Street, Kolkata 700 017
Ashok Rajpath, Patna 800 004
Chowk, Varanasi 221 001

Printed in India
BY JAINENDRA PRAKASH JAIN AT SHRI JAINENDRA PRESS,
A-45 NARAINA, PHASE-I, NEW DELHI 110 028
AND PUBLISHED BY NARENDRA PRAKASH JAIN FOR
MOTILAL BANARSIDASS PUBLISHERS PRIVATE LIMITED,
BUNGALOW ROAD, DELHI 110 007

Contents

DEDICATED TO THE MEMORY & MISSION OF MAHARṢI DAYĀNANDA SARASVATĪ

It was he who proclaimed during the midnineteenth century that the Veda is the Book of all Sciences, and wrote various chapters detailing the Vedic Science of Physics preceding the discoveries made by Modern Science.

Also dedicated to all the big and small names belonging to the scientific community of modern times who contributed and are contributing towards the present development of the Science of Physics which stands today as the living proof of the veracity of the Vedas and Vedic Physics which has always been proclaiming and beaconing that the pursuit of truth must ever be objective, religiously avoiding any pitfalls of creed or belief, howsoever old, fascinating and/or widespread.

DEVANAGARI LETTERS AND THEIR INDO-ROMANIC EQUIVALENTS

with pronunciations exemplified by English words

Vowels

अ	_	a	as in rural
आ	ा	ā	as in tar, father (tār, fāther)
इ	ि	i	as in fill, lily
ई	ी	ī	as in police (polīce)
उ	ु	u	as in full, bush
ऊ	ू	ū	as in rude (rūde)
ऋ	ृ	ṛ	as in merrily (merṛily)
ॠ		ṝ	as in marine (maṛīne)
ऌ		lṛ	as in revelry (revelṛi)
ॡ		lṝ	as in the above prolonged
ए	े	e	as in prey, there
ऐ	ै	ai	as in aisle
ओ	ो	o	as in go, atone
औ	ौ	au	as in Haus (as in German)
अं	.	ṉ	ṅṁ either true anusvāra n or the symbol
अः	:	ḥ	of any nasal symbol called Visarga

Consonants

Equivalents and Pronunciation

क्	k	in seek
ख्	kh	in inkhorn
ग्	g	in gun, get, dog
घ्	gh	in log*h*ut
ङ	ṅ	in sing, king, sink (si*ṅ*k)
च्	c	in dolce (in music)
छ्	ch	in churc*hh*ill (curc*h*ill)
ज्	j	in jet, jump
झ्	jh	in hedgehog (hej*h*og)
ञ	ñ	in singe (si*ñ*ge)
ट्	ṭ	in true (*ṭ*rue)
ठ्	ṭh	in ant*h*ill (anṭ*h*ill)
ड्	ḍ	in drum (*ḍ*rum)
ढ्	ḍh	in redhaired (reḍ*h*aired)
ण्	ṇ	in none (noṇe)
त्	t	in water (as in Ireland)
थ्	th	in nut*h*ook (more dental)
द्	d	in dice (more like t*h* in t*h*is)
ध्	dh	in adhere (but more dental)
न्	n	in not, nut, in
प्	p	in put, sip
फ्	ph	in up*h*ill
ब्	b	in bear, rub
भ्	bh	in ab*h*or
म्	m	in map, jam
य्	y	in yet, loyal
र्	r	in red, year
ल्	l	in lull, lead
		in (sometimes for ड् ḍ in Veda)
		in (sometimes for ढ् ḍh in Veda)
व्	v	in ivy (but like w after cons.)

श्	ś	in sure (śure)
ष्	ṣ	in shun, brush
स्	s	in saint, sin, hiss
ह्	h	in hear, hit

ABBREVIATIONS

A.V.	Atharva Veda
K	Kelvin
K.E.	Kinetic Energy
Nir	Nirukta
Pat. Yog.	Pātañjala Yogadarśana
P.E.	Potential Energy
R.V.	Ṛgveda
Sāma	Sāma Veda
Sāṅkhya/s	Sāṅkhya Darśana
Śatapatha	Śatapatha Brāhmaṇa
Śve. U.	Śvetāśvatara Upaniṣad
Tai	Taittirīya Brāhmaṇa
Vaiśeṣika	Vaiśeṣika Darśana
Y.V.	Yajurveda

FOREWORD

Said al-Andalusi, a noted Arab scholar, stated that "India is the first nation to have cultivated science" and praising Indians for their knowledge further says, "India is known for the wisdom of its people. Over many centuries, all the kings of the past have recognized the ability of the Indians in all branches of knowledge." Referring to theology he writes, "Some of them (Indian people) believe in the creation of the world, while others believe in its eternity...the majority of the Indians believe in the eternity of the world because it is created by the creator of the creators." As regards the cosmology of Indians Andalusi remarks "...(they) say that all the seven planets and their apogees and perigees meet in the head of Aries once every four thousand thousand thousand years and three hundred thousand thousand years and twenty thousand thousand solar years. They call this cycle the 'period of the universe' because they believe that when all the planets meet in the head of the Aries everything found on earth will perish, leaving the lower universe in a state of destruction for a very long time until the planets and their apogees and perigees disperse back to their zodiacs (constellations). When this takes place the world returns to its original state. The cycle repeats itself indefinitely." (*Book of the Category of Nations* by Said al-Andalusi – ed. by Sema an I. Salem and Alok Kumar, University of Texas Press, Austin, Texas, 1991)

It appears that some aspects of Indian cosmology, particularly the cyclic nature of the creation and annihilation, had travelled to the Arab scholars and probably through their writings to the European world. But Indians had a much deeper insight and an equally strong system of correlation of cause and effect in interpreting natural phenomena.

The scientific perceptions of ancient Indian genius are reflected in concepts regarding the ultimate structure of matter, which were first propounded by the Indians. The evolution of elements which are the building blocks for forming diverse compounds has been discussed in various schools of philosophy in India. Ancient Indians had a fairly good understanding of measuring and mapping, of investigating the course of heavenly bodies, of agricultural techniques and of analyzing the constitution of matter. The sources of various scientific perceptions are traced to Ṛgveda. One is simply wonderstruck to find in certain hymns a searching enquiry into the creation of the world. The song of creation is described in the 129th Sūkta of the 10th Book of Ṛgveda which says:

Then was not non-existent nor existent;
There was no realm of air, no sky beyond it.
What covered in, and where? and what gave shelter ?
Was water there, unfathomed depth of water?

It goes on to say that in the beginning one could not say with surety about the immortality or death, nor could one say about day or night. Whatever existed (or did not exist) was void and formless and so on. Then the hymn raises some fundamental questions:

Who verily knows and who declares it,
Whence it was born and whence comes this creation ?
The gods are later than this world's production,
Who knows, then, whence it first came into being ?

Thus Ṛgveda on the one hand raises serious philosophical doubts and on the other has the seeds of a very advanced cosmology. The Vedic cosmos appears to be self-perpetuating and self-sufficient as well. It may thus be recognized that this Vedic hymn does not propose any outside creator. There is no extraneous matter from which 'God' created the Universe.

The history of Indian science is closely linked with the origin of the Vedic texts. According to one view, the Veda not only contains some illuminating truths of scientific knowledge but contains even the truths that physical science has discovered in the modern times. This claim may require a good deal of proving, but it can be said that not only ancient Indian civilization but even several other ancient civilizations possessed secrets of science; some of which modern knowledge has recovered, extended and made more rich and precise but others are even now being recovered.

The language of the Veda is symbolic, and the knowledge that it contains in regard to Matter, Life, Mind and Supermind, as also of unity of the universe and the oneness of Ultimate Reality is presented in a language that is not easily intelligible to us. It speaks of Matter and the knowledge of Matter as that of the three earths; it speaks of life-force as the mid-region (*antarikṣa*), and it speaks of mind and the knowledge of the mind as of three heavens. And beyond the triple lower world of Matter, Life and Mind, it speaks of truth-world, the world of the Truth, Right and Vast (*satyam ṛtam bṛhat*) manifested in *svar*, with its three luminous heavens. The Veda goes still farther, and it expounds the knowledge of the still higher three worlds which in the later tradition of Indian knowledge has been identified with the knowledge of the worlds of the Being, Conscious-Force and Delight. To the Vedic seers all this knowledge was scientific, considering that knowledge is systematic and it is verifiable, repeatable and capable of further expansions in the light of constant enlargement.

In the field of Astronomy the Vedic scholars had scaled amazing heights. The earliest tradition of astronomical observations is perhaps the ancientmost as is evidenced by the specific references in the Vedic and Brāhamaṇical literature. Ṛgveda records the identification of the planet Jupiter (Bṛahaspati) by Vāmadeva. The hymn is also repeated in Atharvaveda.

बृहस्पतिः प्रथमं जायमानो महो ज्योतिषः परमे व्योमन् ।
सप्तस्यास्तु वि जातो र वेण विसप्तरश्मिरधमत् तमांसि ।।

RV - IV.50.4, AV - XX 88.4

Wilson translates this as: "Bṛhaspati when first being born in the highest heaven of supreme light, seven-mouthed, multiform (combined) with sound, and seven-rayed has subdued darkness."

In the Taittirīya Brāhmaṇa one finds a reference:

बृहस्पतिः प्रथमं जायमानस्तिष्यं नक्षत्रमभिसम्बंभूव ।

T.Br. III 1.1.5

Just as Vāmadeva identified the planet Jupiter, the planet Venus (Śukra) was identified by *Ṛṣi* Vena, son of Bhṛgu. The specific reference to this is found in Ṛgveda

अयं वेनश्चोदयत् पृश्निगर्भा ज्योतिर्जरायू रजसो विमाने ।
इममपां संगमे सूर्यस्य शिशुं न विप्रा मतिभी रिहन्ति ॥

R.V.X.123.1

There are two hymns of twenty verses associated with the name of Vena Bhārgava. The planet he identified is named after him. The name Venus appears to have been derived from Vena. The Greek story of Venus as goddess of love also appears as a modification of the Vedic narrative which describes Vena as being loved by an *Apsarā* just as a lady cherishes her paramour.

The Vedic literature indicates a tradition of scientific observation and there are several references which reveal that such a tradition continued in different disciplines.

It would be, therefore, in the fitness of things that serious studies should be undertaken for identifying the scientific concepts in ancient Indian literature and to see whether a tradition of scientific enquiry was built by them and whether they had developed a logic and a consistent approach to understand the ever changing physical world or what is generally called as the mystery of the Universe. In such an investigation the intention should not be to match the Vedic concepts with those of the modern sciences.

The present volume on Vedic Physics by Keshav Dev Verma is indeed a unique attempt to interpret the ancient Indian literature by defining various symbols, concepts and the terminology occurring in Vedic hymns and other texts. While accepting Maharṣi Dayānanda's view that Vedas are the repository of all true sciences the author does examine this statement with a view to test it on the hard rock of truth.

As mentioned earlier, the Vedic hymns contain the seeds of a highly advanced cosmology. Shri Verma has selected Sāṅkhya which according to him 'presents a wholesome structure of Vedic Physics'. It is well known that Sāṅkhya-Pātañjala system explains the physical world (Universe) on the basis of Cosmic evolution; the Vaiśeṣika-

Nyāya expounds the methodology and elaborates the concepts of physics, chemistry and mechanics. Shri Verma has very systematically tried to interpret the Sāṅkhya aphorisms and concludes that the ultimate ground to which the manifested world can be traced is *Prakṛti* having three attributes—*Sattva* (existence), energy at rest, *Rajas* (energy that which is efficient in a phenomenon and is characterised by a tendency to move and overcome any resistance) and *Tamas* (mass or inertia) which resists the *Rajas* to do work and also resists *Sattva* from conscious manifestation.

The ultimate building blocks of the Universe, then according to Sāṅkhya, are (1) Essence, (2) Energy and (3) Matter, mass or inertia. The author concludes, since essence is energy at rest and *Rajas* is energy in motion and mass is energy quantized, the *Prakṛti* is energy. The question now arises as to how does this unmanifested *Prakṛti* become manifest? The Sāṅkhya answers it and says, by coming in contact with the *Puruṣa* - the efficient cause, the inert *Prakṛti* gets activated. The question remains how *Puruṣa* comes in contact with *Prakṛti*? Who initiates this process? Is the process mechanical or non-mechanical? Shri Verma has tried to answer this by taking recourse to Maharṣi Dayānanda's Introduction to the commentary on Ṛgveda and other Vedic hymns. I believe this would motivate scholars for further research.

In one sense Sāṅkhya propounded the theory of evolution rather than creation. The universe is not a creation by an extra-terrestrial agency but the result of the interaction between *Prakṛti* and *Puruṣa*. Historically it is the first doctrine expounded anywhere in the world to exhibit an independence of mind and freedom for enquiry. It is indeed laudable that Shri Verma has made a serious attempt to give a new interpretation to several Vedic statements which reveal this spirit of enquiry. It is a different matter whether one agrees with the interpretation or not but it has to be recognised that the present book stimulates serious interest in the subject and offers a new approach.

Shri Verma has raised a very important issue regarding the important role of the efficient cause which activates the material cause, i.e., the energy to produce this universe as a self-contained whole system. He tries to resolve it by referring to Śvetāśvatara Upaniṣad and defines "It (Efficient Cause) has neither any work to perform nor there are in existence its implements, nor there is anyone its equal, nor is there visible or known to be anyone exceeding It. The excellence of Its power is heard or known from various sources and Its knowledge, force and dynamism are natural states of Its being."

Prakṛti is an inert, unconscious entity while *Puruṣa* is the conscious activator of *Prakṛti*. In Sāṅkhya scheme the evolution of both inanimate and animate or conscious world is taken into account. The present author has not considered the elements related to the biological development; he has only taken into account the evolution of inanimate matter. But the current trend in science is to enquire into questions like, 'How does brain behave as an instrument of mental processes?' The mind-body relationship is a very fascinating yet a baffling riddle and is a subject matter of modern researches. The non-computational aspects of the universe, however, do not find any place in the modern scientific theories. While discussing the incompleteness of such a scientific world view Roger Penrose says:

> "A scientific world-view which does not profoundly come to terms with the problem of conscious minds can have no serious pretensions of completeness. Consciousness is part of our universe, so any physical theory which makes no proper place for it falls fundamentally short of providing a genuine description of the world. I would maintain that there is yet no physical, biological, or computational theory that comes very close to explaining our consciousness and consequent intelligence; but that should not deter us from striving to search for one."

Arguing further, he states:

> "In Part I (of the book) I argued (in the particular case of mathematical understanding) that the phenomenon of *consciousness* can arise only in the presence of some non-computational physical processes taking place in the brain. One must presume, however, that such (putative) non-computational processes would *also* have to be inherent in the action of the same material, satisfying the same physical laws, as are the

inanimate objects of the universe. We must therefore ask two things. First, why is it that the phenomenon of consciousness appears to occur, as far as we know, *only* in (or in relation to) brains — although we should not rule out the possibility that consciousness might be present also in other appropriate physical systems? Second, we must ask how could it be that such a seemingly important (putative) ingredient as non-computational behaviour, presumed to be inherent — potentially, at least — in the actions of all material things, so far has entirely escaped the notice of physicists?" (*Shadows of Mind*, p. 8, 217, Vintage 1995)

However, the need to incorporate non-computational ingredient in the physical theories, is being increasingly felt. Perhaps the present author or some of his associates would undertake this task of providing this missing link.

In the chapter 'Mahat' the author says that according to Sāṅkhya the term *Mahat* is also interpreted as 'buddhi-tattva' or the element of intelligence. Intelligence is not a material product, and according to Verma it reveals the divine wisdom of *Puruṣa*. Later on in the chapter on '*Puruṣa*' he defines *Puruṣa* as the embodiment of *Sat-Cit-Ānanda*. From the epithet *Cit—meaning* the animating principle of life—is derived the word *Cetanā*—that is consciousness. If we accept that, *Puruṣa* completely pervades *Prakṛti* in order to activate it, then intelligence and consciousness become inherent properties of the manifested world. What Roger Penrose is seeking becomes a natural outcome of the Sāṁkhya system.

But the question still remains as to how *Puruṣa* activates *Prakṛti*? It may be recognized that there has been a debate among the various commentators of Upaniṣads and the Sāmkhya school regarding the ultimate cause of creation. The debate has travelled into modern physics as well. Studies like the present one undertaken by K.D. Verma can be useful in throwing light on this subject by the paradigm suggested by ancient Indian seers and sages.

K.D. Verma's exposition of various concepts like tanmātra, pañca mahābhūta, ākāśa, vāyu, agni, etc., deserves special attention and critical study as also his exposition of Brahmā and Prajāpatis. He has attempted to build a theory of creation by analyzing various hymns of Vedas and by attributing meaning to various Vedic symbols and cosmological processes which provide a holistic world-view converging to a similar approach propounded by some of the modern physicists. However, Verma has introduced some more concepts like Brahmā (not Brahma) as the smallest creative principle termed as 'Atom', Prajāpatis—ten in number and correlated to various fundamental particles like electron, proton, neutron, pion, etc. Such a correlation is likely to raise a serious debate and requires more research in Vedic philosophico-scientific concepts and terminology and for demystifying the Vedic literature. Verma's book will serve as a catalyst for further research in developing a consistent theory based on Vedic concepts for explaining the various properties and behaviour of the physical world around us. Vedic scholars should undertake this task seriously.

K.D. Verma has based his model mostly on the basis of *Sāṅkhya* philosophy. But there are other schools of Hindu thought which hold an atomistic view of the physical world around us. It is now well recognized that the Atomic theory makes an integral part of the Vaiśeṣika, and it is also acknowledged by the Nyāya. Jains have also adopted the atomic theory, as is stated in the *Abhidharmakośavyākhyā*. According to some scholars, although no mention is made of it in the Buddhist Pāli canonical books, it is quite different, however, with the Northern Buddhists. The Vaibhāṣikas and Sautrāntikas were adherents of the atomic theory, while the Mādhyamikas and Yogācāras opposed it, as they declared the external world not to be real.

It appears that the Jains worked out their system from the most primitive notions about matter. According to some commentators the Jains maintain that everything in this world, except souls and mere space, is produced from matter (*pudgala*), and that all matter consists of atoms (*paramāṇu*). Each atom occupies one point (*pradeśa*) of space.

Vaiśeṣika mainly deals with physics, and Nyāya with metaphysics and dialectics. The physical side of the atomic theory was deliberated more by the Vaiśeṣikas, and the metaphysical by the Naiyāyikas. In fact, Bādarāyaṇa regards the atomic theory as the cardinal principle of the Vaiśeṣika system. The *Nyāya Vārtika* states that atom is invisible because it is not composed of material parts.

The metaphysical questions, however, relating to atoms are fully discussed by Gautama, and further explained by Vātsyāyana. The Naiyāyikas maintain that the whole is not merely a combination of its parts but something more than its parts; it is a different thing *(arthāntara)*, not separated from its parts, but rather something in addition to them.

In the Buddhist thought the Vaibhāṣikas admitted that an atom had six sides, but they maintained that they made but one, or what comes to the same, that the space within an atom could not be divided.

The Sautrāntikas seem to have regarded the aggregate of seven atoms as the smallest compound *(aṇu)*. Their opinion seems to have been that the (globular) atoms did not touch one another completely, but that there was an interval between them. *All agreed that atom is indivisible*, though some admitted that it might be regarded as having parts, viz., six or eight sides. Both Vaibhāṣikas and Sautrāntikas declare that atoms are not hollow, and cannot penetrate one another.

The latest improvement of the atomic theory consists in the assumption of *dvyaṇukas,* etc. It was first thought by Praśastapāda and is plainly referred to by Udyotakara; it was received as a tenet in all later works of what may be called the combined Nyāya-Vaiśeṣika.

It is assumed that two atoms *(paramāṇu)* form one binary *(dvyaṇuka)*, and that three or more *dvyaṇuka-s* form one *tryaṇuka,* which is 'great' and perceptible by the eye. From *tryaṇukas* are produced all things. Modern writers further assume *caturaṇukas*, formed of four *tryaṇukas*, etc. Thus the molecular structure of matter is well recognized by the followers of the Nyāya-Vaiśeṣika system.

The idea of the infinitesimal in this sense seems to have already been current in the time of the Upaniṣads, where we frequently meet with the statement that *Brahman* is subtler than the subtle, and larger than the largest and that the self *(ātman)* is small *(aṇu)*. In order to arrive at the conception of the atom, the idea of the infinitesimal had not only to be applied to matter, but it had, at the same time, to be joined to the idea of its indestructibility. On the other hand the Buddhist view was not of eternal atoms, for they considered *Saṁsāra* as continual springing into existence and annihilation and hence the whole physical world was nothing but an aggregate of non-eternal atoms.

Thus one finds that, there were several schools in the ancient Indian physics and metaphysics where the nature and properties of gross matter were interpreted in terms of the atomist view of nature. However, Shri Verma has proposed an entirely different picture of atom and other fundamental particles. He has identified atom with the 'Brahmā' of the Vedic text. Once again this would inspire serious research in the Indian tradition of scientific perceptions—particularly in physics.

Before concluding, I would like to make some remarks about the chapter on *kāla* (Time). The author has established that the Vedic seers on the one hand describe the smallest unit of time as one millionth of the second and on the other, talk about the vast span of time extending to billions and billions of years. The cyclic nature of creation and dissolution does not agree with the linear arrow of time. Further, the issue of the 'beginning' of time and its nature remains an enigma even today. The author has raised this question by asking what happens to time when dissolution occurs? Does time also get dissolved? The Sāṁkhya has not enumerated 'time' in its list of 25 categories and is silent regarding the role of time in initiating the process.

But Shri Verma argues that the 129th Sūkta commonly known as 'Nāsadīya Sūkta' of the 10th Maṇḍala of the Ṛgveda is to be interpreted to denote the existence of *kāla* even when sat and asat were non-existent. In support, he quotes a paper read by Dr. Mahavir at a seminar held at Ajmer in October 2000. The above Sūkta begins as *nāsadāsīnno sadāsīttadānīm...* Dr Mahavir has interpreted this part of the Sūkta as at that time (*tadānīm*) neither *sat* nor *asat* were in existence, which according to him is a positive statement regarding the existence of time during the state of dissolution. The Vedic scholars have so far not paid their attention to the scientific explanation of all of Bhāvavṛttam sūktas and that of this particular mantra. The existence of time during the state of dissolution and the calculation of time during that state, which is a serious scientific challenge, has been dealt with at length in Vaiśeṣika and other Indian *darśana* books.

Verma has further pointed that Vaiśeṣika treatise states kāla (time) to be eternal. During the period of *pralaya* or dissolution it remains dormant and re-appears in a state that is measurable during the period of creation, for it provides the basis for measuring all sorts of activities let loose in the realm of relativity, and also serves as an instrument to human beings or any such form of highly intelligent species emerging therein to learn and verify the divine knowledge.

It is interesting to note that the Atharva Veda has several hymns in praise of Time. The verses in the XIXth Book of the Atharva Veda are perhaps unparalleled pieces of poetry in any literature and also illustrate the depth of the comprehension of Time by the Hindu mind. The Vedic seer in the Atharva Veda sings in praise of Time as translated by Swāmī Satya Prakāśa Sarasvatī:

> In time fervour, in Time the great observer, the highest knowledge is well set. Time is the Lord of all; father of the Lord of creatures is He.
>
> A. V. XIX 53.8

> Urged by Him (Time) created by Him; all that is set, surely, within Him. Time becoming the Divine supreme (Brahmā) sustains the Lord seated in the highest abode.
>
> A. V. XIX 53.9

> Time has created the creatures; in the beginning, Time created the Lord of creatures. The self-existent seer (Kaśyapa) is born from Time; from Time fervour is born.
>
> A. V. XIX 53.10

> By Time, the son (of Time; Prajāpati) created past, present and future in ancient times. The *ṛcās* (Praise-verses) are born from Time; from Time is born the Yajuḥ (sacrificial-text)
>
> A. V. XIX 54.4

These verses and several references in Purāṇas reveal that the ancient Hindu seers devoted sufficient time and energy for comprehending the true nature of Time.

The Yoga-Vāsiṣṭha has also dealt with the ideas regarding time and space. It consists of 6 Books where the sixth Book itself has two parts.

According to the sixth Book of the Yoga-Vāsiṣṭha (the figures in the bracket refer to the Book, Section and Verse.—'Time' cannot be analyzed; for however much it is divided it survives, indestructible [1.23]. There is another aspect of this time, the end of action (*kṛtānta*), according to the law of nature (*niyati*) [1.25.6-7]. The world is like a potter's wheel: the wheel looks as if it stands still, though it revolves at a terrific speed [1.27]. Just as space does not have a fixed span, time does not have a fixed span either. Just as the world and its creation are mere appearances, a moment and an epoch are also imaginary [3.20]. Infinite consciousness held in itself the notion of a unit of time equal to one-millionth of the twinkling of an eye; and from this evolved the time-scale right upto an epoch consisting of several successive sets of the four ages, which is the life span of one cosmic creation. Infinite consciousness itself is uninvolved in these, for it is devoid of rising and setting (which are essential to all time-scales), and it is devoid of beginning, middle and end [3.61].

There are several discussions related to the concept of Time in different schools of ancient Indian philosophy. The Mīmāṁsā school apart from confirming the Vedic view also considers that time gives an idea of decay. The tendency of time to cause decay is regarded as a characteristic feature of time in Mīmāṁsā. This characterisation has an unmistakable resemblance to the thermo-dynamic arrow of time given by the entropy law, since increase of entropy has a standard interpretation of progress towards chaos, or a decay of order. (C.K. Raju. "Time in Indian and Western Tradition: in Mathematics, Astronomy and Biology in Indian Tradition" Project of History of Indian Science, Philosophy and Culture, 1995).

The Advaita Vedānta distinguishes between apparent time and real time. Apparent time denotes change, e.g., rising and setting of sun, growth of a tree etc., and is therefore illusory. The Advaita believes that at the fundamental level nothing really changes and hence real or absolute time cannot be conceived. The Jain view of time is atomic, while Buddhist concept of time as instant takes one to the notion of momentariness where past and future do not exist. The cyclic, linear and spiral nature of time all have been discussed by different schools of Indian thought. Shri Verma has attempted to underscore that it is the ancient Indian wisdom which has deliberated on this very complex issue of Time and attempted to solve the riddle of Time which though appears to be ubiquitous is yet difficult to comprehend. Shri Verma's book undoubtedly impels the serious-minded scholars to apply their mind for solving the riddle of time.

From the foregoing one can easily see that while on the one hand the ancient Indian genius had been deeply involved in understanding the mystery of the origin of the universe on the other various schools had tried to analyse the structure of gross matter based on an atomistic view of Nature. One can also recognise that on both the physical and the metaphysical levels there exists a continuing tradition ever since the Vedic times. It should be the endeavour of Indian scholars to develop a consistent approach to the world view based on one or the other school. Shri Verma has attempted one such exercise based on the Vedic Model which needs to be further extended. This work will certainly inspire other serious-minded scholars to undertake further research on this count and provide a deeper understanding of 'Atom' and the 'Universe'. I congratulate Sri Verma for producing this book.

(MURLI MANOHAR JOSHI)

06-06-06
6, Raisina Road
New Delhi

FOUNDER EDITOR'S EXORDIUM

When the Management of Motilal Banarsidass accepted my idea of a two-pronged Indology series on India's Scientific and Cultural Heritage as a part of the Centenary celebration of the famed Publishing House with its characteristic corporate generosity of heart, I was also conscripted to be the General Editor of the twin series. The idea of the series was in line with the sustained and manifold contributions of this preeminent publishing house to Indological studies over an entire century. Those contributions have put scholars in their abiding debt. I agreed to be the General Editor with a sense of hesitant trepidation and despite my feeling that I was unequal to the task, mainly because it is my deep conviction that every successive generation of Indians owes to itself the debt and duty of renewed understanding and renascent and revitalized reiteration of their heritage in which the scientific, cultural and spiritual have naturally, uniquely and creatively converged.

India's Scientific Heritage Series is particularly intended to illumine the path, the quest and the achievements of the world's oldest scientific literature and tradition in India and to create an objective understanding of India's history of scientific thought which was an integral part of an unfragmented and holistic understanding of the Universe and the *Homo Sapiens.* This Scientific Heritage series is meant to be a humble but significant contribution to the world history of science and philosophy of science. I believe there is a deep pfelt need for a full-fledged multidisciplinary discipline of History of Indian Science and Technology and in that context, I applaud the initiative of the Infinity Foundation in U.S. led by Dr. Rajiv Malhotra. I hope the idea of the discipline of History of Indian Science will materialise in our time. It is noteworthy that Samuel Huntington had pointed out in his book, *Clash of Civilizations,* that in 1750 China accounted for almost one-third, India for almost one-quarter and the rest for less than a fifth of the world's manufacturing output. I take my cue and a measure of comfort from Professor Amartya Sen's reference (1997) in the Indian scientific context to Dr. Joseph Needham's 30+ volumes on Chinese history of science and technology which is sure to make a lasting and far-reaching impact in assigning to China its place of pride and honour in the world history of science and technology. Says Amartya Sen, whom I quote (without his side swipe on Vedic Mathematics):

> "Fear of elitism did not, happily, deter Joseph Needham from writing his authoritative account of the history of science and technology in China, and to dismiss that work as elitist history would be a serious neglect of China's past... A similar history of India's science and technology has not yet been attempted, though many of the elements have been well discussed in particular studies. The absence of a general study like Needham's is influenced by an attitudinal dichotomy. On the one hand those who take a rather spiritual, even perhaps a religious, view of India's history do not have a great interest in the analytical and scientific part of India's past.., missing the really creative period in Indian mathematics by many centuries. On the other hand, many who oppose religious and communal politics are particularly suspicious of what may even look like a 'glorification' of India's past. **The need for a work like Needham's has remained unmet.**"

Truly notable were ancient and medieval India's scientific and technological contributions in theoretical Sciences, including linguistics, mathematics, astronomy, cosmology, cosmogony, physics, metaphysics as well as life science and technologies of sculpture, water harvesting, textiles, shipping, ship building and weaponry, not to speak of the advances in the science of law, governance, international relations and sustenance of a value-based society. Misguided and superficial secularists forget in their frenzy that those were the foundations of a national and secular outlook in ancient Indian Society and lapse from which paved the way for India's decline and defeat. Tragically, apologists of that vision of secularism and their camp followers have little sense of India or of history and no pride in India's past heritage.

I am thankful to Professor Murli Manohar Joshi for contributing at my instance an erudite Foreword rooted in the intertwined philosophy of India, physics and metaphysics. His call for a debate on the exposition and perspective of the author of the present book will, I hope, evoke deeper and wider scholarly discussion in the discourse of India's scientific heritage.

Before Needham and his collaborators attempted their *magnum opus*, the prevailing Western notion of Chinese science was as observed by Whitehead:

> "There is no reason to doubt the intrinsic capacity of individual Chinamen for pursuit of science. And yet Chinese science is practically negligible. There is no reason to believe that China, if left to itself, would have ever produced any progress in science. The same may be said of India."

The ghost of the weird notions about Chinese science conjured up by Whitehead is being laid to rest, but the western ethnocentric stereotype about India continues to be perpetuated. For instance even Toynbee, who knew Indian civilization better, wrote:

> "A mechanical penchant is as characteristic of the western civilization as the aesthetic penchant was of the Hellenic, or a religious penchant was of the Indian and the Hindu."

It is time for modern and resurgent India to wake up and exorcise the ghosts of Western obscurantism rooted in studied ignorance of India's Scientific Heritage and to make the world appreciate objectively the scientific temper and mindset which created a creative communion of rationality and spirituality that indeed is the *raison d'etre* of the Indian Scientific Heritage Series.

L.M. Singhvi
Founder Editor

INTRODUCTION

One, who goes through the book, is sure to exclaim, Remarkable! Marvellous! notwithstanding an apparently polar tradition of the Vedas and the modern science, the subject of the book is undoubtedly remarkable and wonderfully dealt with. Rarely indeed one comes across such a grand exposition of the Vedas, of Indian scientific and philosophic traditions in the light of modern science, that reflects the comprehensive grip over both the disciplines, singly and jointly, on the part of the author. The natural outcome is a step ahead towards cherished Grand Unification Theories (GUT) like that of Quantum Field Theory due to clues and insight especially from Vedic tradition. It also is an attempt of uniting the East and the West that are also diametrically opposite to each other. But at deep, fundamental level or at the top transcendental level, we see that this opposition gets dissolved or gets evaporated. However these two extremes may appear to be two poles apart like the two faces of the Greeko-Roman divinity Janus or the Vedic twin divinities of Aśvinī Kumāras, the Physicians of Gods.

The Vedas stand for self-revelation, cosmic revelation, self-realization and cosmic realization – an inner journey towards the Reality, the Truth. Modern science follows external journey towards the same. The Vedas, however, are subjective, experiential, continuous, analogous, qualitative, supra-sensory and supra-rational yet direct the field of Spiritual Energy. Science is materialistic, objective, experimental, digital, logical, rational with its source in axioms, postulates, basic principles and laws of Nature – the field of Matter and Energy.

The knowledge gained by Vedic seers is called *Yogaja-pratyakṣa,* that of scientists is called *Indriya pratyakṣa,* i.e., both are direct perceptions (*pratyakṣa* = directly through eyes, i.e., sense faculty in general), but belong to different orders. The one is a sudden explosion and the other is step-by-step exposition. The one implies the other, the personal and impersonal aspects of the same reality. Both could mutually conform or confirm; the unification or integration of both would be the cherished height of human profile.

The Sāṅkhya, on the other hand, is the bridge in Indian context between Vedic or Yogic perception, the scientific traditions as well, and also between the Eastern and the Western epistemologies. It has its source and base in the principle of duality, primarily binary, numerical (*saṅkhyā* = number, while Sāṅkhya as a school implies proper expression which is quantitative and numerical as well) knowledge representation.

The major thrust of modern science, not to exclude the classical and pre-classical one, is the theory of creation of the universe, its beginning (Big Bang), sustenance and dissolution (the big crunch) – the evolution. It addresses *inter alia*, the four natural forces, the world of matter and anti-matter, the microcosmic (Quantum mechanics), the macro-cosmic (General Relativity) including space and time. The Vedas and Sāṅkhya (i.e., Indian philosophical and scientific tradition as a whole, notwithstanding internal differences and distinctions) also centre around the same issue. However, evolution in Indian context also anticipates involution (cyclicity). Sāṅkhya, within its duality or binary algorithm floats total twenty-five principles scheme, wherein *Puruṣa* (Cosmic person) is one, the other is *Prakṛti* (Creatrix) which contains within twenty-four principles with eight major and sixteen secondary, evolved out of these eight. As compared to *Prakṛti* as One, *Puruṣa* is also One, but in the process of evolution. He inspires *Prakṛti* to transform itself into a cosmic system. *Prakṛti* comprises *Mahat* (lit. the Great One), Ahaṅkāra (individuality/ego), three *guṇas* (qualities/properties), five subtle elements principles (called *pañca tanmātras*) like sound, touch, colour, taste, smell, five gross elements (pañca-mahābhūtas), which correspond to *Ākāśa* (ether), Air, Fire, Water and Earth, respectively to the former group. The present author includes some more concepts, principles or categories from Vedic lore like ṛitam (dynamic principle), Satyam (essential, existential principle), Brahmā (Atom, the smallest creative principle, which is all-pervading), Prajāpati (the Lords of creation ten in number), yajña (cosmic sacrifice as well as the smallest version of it), kāla (Time), pralaya (dissolution), mahat sphoṭa (the Big Bang). In principle this scheme covers implicitly the concepts, the theories of Creation from pre-classical, classical and modern science as well.

In fact, as the title itself implies, there have been three major drives in this direction of inter-action between ancient Indian cultural tradition and the western thought that goes back a few centuries starting from Europe. The first is that of derogatory approach to condemn the Indian tradition in every respect, as against the Western, especially the modern materialistic one. The other is the positive one which discovers that there had been no antagonism between religion (church), philosophy and science (mind and matter, science and spirituality) conspicuously present in the West, but absent in Indian tradition. The third is the one that is due to the very rich and profound Vedic and Yogic heritage followed by philosophical and scientific tradition that tries to find out the clues and get insights from it, then to unify not only science in general, but the two most powerful, seemingly contradictory and competitive parallel theories of Quantum Mechanics and General Relativity, that have been desperately cherished and tried out by no less figures than Max Planck, Einstein, Niels Bohr, Schrodinger, Heisenberg, David Baum, not to speak of Abdus Salam, Penrose, Hawking etc. In India Swāmī Dayānanda Sarasvatī was the first powerful person followed by Swāmī Vivekānanda, then by Maharṣi Aurobindo, equally powerful figures, attested by their works, available in print. The second and the third drives also were addressed by them. Their lives and activities were examples of this spirit of unification.

The second and the third drive is found reflected in the works of the Indologists in the West, and numerous other Western scholars, researchers and scientists during last few decades. Such prominent works are; *Tao of Physics, The Dancing Wu Lie Masters, Physics as Metaphor, Brahman is equal to Mc²*, etc. Therein, an attempt has been made to see parallels between ancient Indian tradition (Vedic/Buddhist/ Jain), philosophical, scientific and others, and Western tradition, both religious and scientific. Sincere effort is made therein to get some light, some glimpses, some breakthrough to overcome the antagonism between science and religion, spirituality and materialism, matter and consciousness, and at the end a royal road to unify, integrate the Quantum Mechanics and General Relativity. A number of Indian scholars too have started treading on this path.

In the post-Independence era there have been many strides in this direction, most powerful of which have been that of Maharṣi Mahesh Yogi and his team of scientists, Swami Raṅganāthānanda of Ramakrishna Mission, Swami Satyaprakasha, Bhagavaddatta, ISKCON group and recently Prajāpitā Bramha Kumārī Group. For Shri Keshav Dev Verma the source of inspiration is Maharṣi Dayānanda Sarasvatī, who stood against the committed serious attempts to deride Veda undertaken by a few Western scholars and those Indologists who followed them blindly. His lineage was continued by some of his followers. With this zeal and inspiration, keeping pace with those who are engaged in Grand Unification, has come this work. Shri Verma's mission, all this, is the result of his last thirty-five years' efforts.

Modern science actually begins with Galileo, Copernicus and Kepler, grows with Newton, matures, transcends and reaches height with Max Planck, Einstein, Niels Bohr, Schrodinger, Heisenberg and others. Historically from Newton onwards, the Nature, the Reality, were encapsulated in formulae, equations, derivations, and were thus brought down on the slate and pencil, subject to various transformations in the hands of the scientists.

Till then Indian scientific tradition, that followed the Vedas, was ahead in almost every field, the world over, and the Vedas did reflect definitely the then sciences and scientific tradition. There are some problems, however, that are posed while understanding and translating them due rather to their being very much ancient, couched in most ancient version of Sanskrit, loss of continuity and communication, the very cryptic, mystic and codified form of Vedic lore itself. For those who are not initiated into it, nor exposed to the tradition, this is very much the truth. In case of modern science, excepting those who are not directly in it, similar is the case.

Science accepts observer, a witness, albeit outside the lab, outside the office, the realm of science, built up by him only. Neither can it locate him nor can accept *a priori*, as India did long back in the past by positive and negative (*anvaya* and *vyatireka siddhi*) epistemological deliberations as consciousness, the very property of the soul, that is concomitant with existence and bliss. That is the true nature of the soul. It turns out to be the self, the soul. The Quantum Mechanics, notwithstanding this, has let it enter inside its kernel, as something more than the observer, the participant, the very component of the experiment. Sāṅkhya following Vedas accepted *Prakṛti* (the Supreme Creatrix) as the material cause, and *Puruṣa*, (the Supreme Person) as the efficient cause of the universe, thereby forming the link, the bridge between Vedas, Indian scientific tradition, as well as modern science in general.

K.D. Verma has demystified the Vedic hymns by decoding the symbols, technical terms and concepts, thereby many of the mute problems, some fundamental issues in Indian scientific tradition in general and in particular mathematics, astronomy, physics, material and life sciences, have been resolved, interpreted properly, and this is the most positive contribution of the present author. Many other occult, esoteric, enigmatic, obscure ideas, concepts and terms in Indian culture have been made intelligible, meaningful, significant, and relevant, even scientific by the author for the first time, so lucidly. This being highly commendable, puts him in the list of illustrious predecessors.

With this background one can appreciate the author's venture, when he retrospectively interprets Vedic and Sāṅkhya tradition in the light of modern science. Contextually he then envisages five basic principles (*pañca tattvas* or *pañca tanmātras*) associated further with their respective evolutes, the five gross, structural elements (called *pañca mahābhūtas*), sound and ether, touch and air, form and fire, taste and water, smell and earth (*śabda-ākāśa, sparśa-vāyu. rūpa-tejas, rasa-jala, gandha-pṛthivī*) inclusive not exclusive. They address themselves by their permutations and combinations at various levels. To the author the first group satisfies Quantum state, the microcosm, the other satisfies the General Relativity state, the macrocosm. This is simply stupendous exposition in the book.

His exposition of *Prakṛti* (the material cause)—the field of matter, and of Puruṣa (the efficient cause)—the field of consciousness, the witness (the observer), the Brahmā, the atom, the symbol of smallest and biggest form of Reality, that of Prajāpati (ten in number) of the Vedic pantheon, is superb. Equally illuminating is the exposition of *ṛtam,* satyam, the dynamic and static, kinetic and potential character of the Creation, the Reality, and is no less provoking. Similarly, the interpretation of the cluster of stars, *Saptarṣis,* with their female counterparts in terms of matter and anti-matter world (the phenomena of positive and negative sub-elementary particles) is very much revealing. No less interesting is the explanation of *yajña* (the sacrificial institution), of kāla (time, the comprehensive one), mahat-sphoṭa (vis-á-vis Big Bang) and of pralaya (final dissolution- the Big Crunch).

The most salient feat of this exposition and interpretation is that he has thrown a new light, the unprecedented one, on those and other vital ancient Indian issues which, at the same time, support, complement or confirm the modern scientific discoveries and also advance the endeavours of the scientists' communities, at their initial (the fundamental) and at the top (the metaphysical/philosophical levels) not certainly at the middle (the objective) rational, the field proper and unique to modern science, wherein both, at lower and upper levels we find limits and scopes, of the ancient and modern physical (science) and metaphysical (the philosophical) tenets of the Western and the Eastern traditions (i.e., the expression of the unity and the diversity principle).

The most outstanding outcome of the work by Shri Verma is the Vedic model, Vedic calendar, expressed in mathematical terms, formulated as an equation, that is at Quantum Mechanics and General Relativity and their unity. It is a Unified Algorithm and Unified Equation. For scientists, the Indologists and researchers, the book would be a source of great inspiration.

Dr. S.N. Bhavsar
Pune

PREFACE

Many of the sages of Indian origin have said that Vedas are 'The Truth' and my humble submission is that we should regard the Vedas as the Books of Science. Being regarded as 'The Truth' they are always open for scientific discussion and investigation. In this way, from times immemorial, the Vedas have been open to all sorts of investigation and their doors have always been ajar for all seekers of 'The Truth'. They have served the society and the human beings by revealing 'The Truth'.

In this script, I have based my efforts on the treatise named Sāṅkhya. Sāṅkhya has named ten branches to cover the entire physical world. Five of them cover the field of quantum or micro objects and the rest of the five classical or macro objects. Sāṅkhya says that as a chariot needs a charioteer, so is the chariot of creation launched by the charioteer *Puruṣa*. According to the Vedas, *Puruṣa* overextends or transcends far beyond the realm of ten branches of Sāṅkhya. Yajurveda devotes its entire 31st chapter to *Puruṣa*. The present script deals with this concept in detail. In my opinion *Puruṣa* is the synonym of the Supreme Master of Sciences. The very first mantra of the said chapter of Yajurveda says that the existence of *Puruṣa* covers not only the entire cosmos but transcends far beyond the jurisdiction of the ten branches of governance. Further the third mantra says that the entire realm of creation is covered in Its one step while that of the energy is in Its three steps. Also, *Puruṣa* pervades far more beyond in all directions. From the present script I quote a mantra of Yajurveda as follows:

"sa paryagāt chukramakāyam avraṇa-
masnāviram śuddhamapāpaviddham
kavirmanīṣī paribhūḥ svayambhūr-
yāthātathyatorthān vyadadhācchāśvatībhyaḥ samābhyaḥ"

(Y.V. 40/8)

The Mantra says: *(sa)* That Being *(pari+agāt)* pervades and surrounds the entire universe, *(śukram)* is resplendent, *(akāyam)* formless, *(avraṇam)* unscarred, without cuts, *(asnāviram)* without sinews, *(śuddham)* pure, *(apāpaviddham)* unpierced by any evil like matter antimatter which annihilate each other, *(kaviḥ)* omniscient, *(manīṣī)* and of profound wisdom, *(paribhūḥ)* pervading all around, *(svayambhūḥ)* self-existing, *(yāthātathyataḥ arthān vyadadhāt)* has imbued all products from micro to macro with properties and actions in accordance with the rules of creation *(śāśvatībhyaḥ samābhyaḥ)* for eternal years.

According to the Sanskrit lexicons *śukram* also denotes *brahmāgni* which literally fits to mean absolute zero temperature pervading all over. The next noun *akāyam* speaks of no geometry, no density and *ipso facto* utter vacuum. Other nouns specify the attributes of the state of Being beyond the laws and rules of the quantum as well as classical physics.

If we glance at the modern scientific point of view of the universe it says that matter weighs very little and cosmic space is relatively very much larger than the space occupied by matter. The overall situation is also like a vacuum of infinite dimensions. Quantum Mechanics also prefers that real particles are formed from virtual transitions. If one concludes that the initial conditions of the universe were as if there was infinite eternal vacuum then such a physical state, according to physical laws of gases, leads to a state of temperature equal to absolute zero degree or -273.16^0 Celsius.

Thus, the Vedic vision, which is much older than the age of modern available version of gas laws, seems to be talking about the similar physical state, that is, the original state of universe had particles which had no geometrical size, no matter density and all existing at temperature equal to absolute zero. It should be a perfect 'idle state' with no motion or impulse. This gives birth to a contradicting question with Vedic point of view that there is a charioteer *Puruṣa* who might generate impulse or an 'inner impulse' of the virtual particles. This book sheds light on very fundamental questions such as above. What generates an impulse that becomes cause of real particles ultimately is the central question. Shall we still believe $E=mc^2$? Or there is some other way of finding out total energy in such an 'idle state' which may not be called a state of matter.

Recently, I happened to go through a book entitled Q is for Quantum Particle Physics from A to Z by John Gribbin (University Press India Ltd.) and I agree with his comment on quantum interpretation: "You can even favour one interpretation on weekdays and a different one at the weekend. But the one thing you must not do is to believe that any quantum interpretation is 'The Truth'. They are all simply crutches for our limited human imagination, ways for us to come to grips with the weirdness of the quantum world, which never goes away and is outside the scope of everyday experience."

Like other seers and sages I may also say that Vedas are said to be 'The Truth' but still there is scope for the quest for 'The Ultimate Truth' and Vedas can still guide the human race.

Thus, in a physical state possessing infinite amount of energy without existence of matter and temperature, how may the impulse be generated, whether Einstein's equation of energy or even Planck's constant are valid for such a state, automatically comes under the category of rethink about the Quantum Physics.

2 Ka 15, Kamla Nehru Nagar **Keshav Dev Verma**
Ajmer Road, Jaipur-302024 (India)
Dipawali, the 1st October. 2005
(International Year of Physics)

ACKNOWLEDGEMENTS

I am highly grateful to Dr. Murli Manohar Joshi (MP), former Head of the Department, Physics, Allahabad University, for his erudite essay as a Foreword to this script. I express the same feeling for Dr. L.M. Singhvi, Founder Editor of the India's Scientific Heritage series under which this publication is appearing, who stubbornly held back his clearance till the receipt of Dr. Joshi's contribution. I am no less grateful to Dr. S.N. Bhavsar, formerly of Pune University, for his evaluation report on my work, which, with his kind permission, has been included as Introduction. Dr.Vijay Bhatkar and Dr. B.R. Kulkarni, Director Chemical Laboratory, both well known physicists from Pune, also appear fully covered in the focus of this whole-hearted expression of my feelings for their valuable comments.

I gratefully acknowledge the contribution of Late Prof. Pratap Singhji Rathor, former Head of the Department, Physics, in the Jodhpur University and Bikaner University successively, whom I met when he had completed a couple of years of the ninth decade of his life. Through a daily discussion of about ninety minutes, continuously for over two years, with neither Sunday nor festival breaks, he virtually churned this modern science and taught me physics.

Hearty thanks are no less due to Dr. Vinod Kumar Verma, Associate Professor of Physics at Rajasthan University, Jaipur presently, who always found time for me whenever I desired to discuss some aspect of a mythological anecdote to show that the ancients in India preferred that as a style of narrating scientific events. He even helped me in the choice of the language and the manner of my interpretation. His help has been available to me for a number of years and continues as such.

Thanks are also due to the 'Sharma Trio', namely, Anand Sharma, Ashok Sharma and Y.D. Sharma, who lent their support and help in numerous ways as I traversed this terrain.

PROLOGUE

In India from the ancient times *padārtha-vidyā* or the Science of Physics has been known to be the most prominent subject of study. It is claimed to have its origin, like so many other subjects, in Veda. Vedas are renowned to be the revealed knowledge, and a whole host of the greatest among the scholars have affirmed such a reputation. Although the famous treatise in Sanskrit language known as Sāṅkhya enumerates 25 categories which are instrumental in the creation of matter and universe, yet it got more identified traditionally with five elementary causatives which are known as the five *mahābhūtas* and are included among the said twenty-five elements.

With the dispersal of knowledge from India to many other countries in course of time the five mahābhūtas also travelled abroad in the form of translations of their nomenclature, both eastward and westward. In the west they went as far as Greece which was also a famous seat of learning. It appears that Greek scholars who were famous for their critical study of various subjects, could not appreciate the import of the first one of the set of five, and preferred to ignore the same. But they accepted the remaining four as fundamental elements of creation and from there onwards the translated names spread all over Europe, the West Asian countries and the American continents.

The original Vedic names of the five are ākāśa, vāyu, agni, jala and *pṛthivī*. The first one, as said earlier, was ignored and the remaining four survived in the form of their literal translations in respective languages, which in English language came to rest as Air, Fire, Water and Earth or Soil, and continued to be mentioned as the fundamental elements. Even in India, the land of their origin, the five Vedic nouns mentioned above, due to a communication gap in the study of Veda, came to be associated with their respective conventional and traditional meanings. As the nouns denoted certain material forms the same were mistaken as the fundamental elements.

In the year 1824 of the Christian Era, a boy was born in Gujarat, presently a state in the Indian Union, who grew to be one of the greatest scholars of Veda in modern age and became known as Maharṣi Dayānanda Sarasvatī. He was one of the greatest social and religious reformers. He declared that Veda is the book of all true sciences. He emphatically maintained that in ancient India there were aircrafts in use. The west at that time was seriously attempting to fly, but the heavier-than-air flying was still four decades away. He also wrote about many other scientific inventions of the day made in the west, including the weaponry, or their technique having been mentioned in Veda.

During that period a great campaign was being carried on in India for proselytizing Indians to the Christian faith which enjoyed full approval and support of the British rulers. Maharṣi Dayānanda proved to be the greatest stumbling block for both the Christian as well as the Islamic proselytization efforts. A very serious attempt to deride Veda was undertaken and a few British scholars devoted themselves to the study of Vedic and other relevant texts with a view to present the same in a shoddy light. The mantras were declared to be the "songs of the shepherds".

But Maharṣi Dayānanda stood his ground. His erudition and knowledge knew no bounds, even though he had no benefit of the English language. In all scholarly discussions he bested his opponents. He was the first person to speak and write about a total political freedom from the British rule. His enthusiam created a wave of awakening in India.

The British rule lasted about six and a half decades after the final departure of Maharṣi Dayānanda. No attempt to encourage any scientific research based upon Vedic texts was expected to be made during those years. But the pity is that even after an independent India having entered the later half century of its freedom any association of education with Veda still draws vehement hue and cry and political flak.

Under such circumstances it occurred to test the claim of Maharṣi Dayānanda that Veda is the book of all true sciences. In our opinion such a test ought to be quite comprehensive. We chose the subject of Modern Physics in the terms of which we thought of examining the veracity of the Vedic Physics. The Science of Physics is said to be the science of all sciences and if Vedic Physics can hold itself vis-a-vis the modern, highly developed, and claimed to be so well proven Science of Physics, not only India but also the entire scientific community all over the world, would give a little more respectful consideration to Veda in the field of scientific research. The result of such an endeavour is in your hands. But before you take this case for assessing whatever its worth, we feel it necessary to lay down before you here, the handicaps and limitations of this exercise.

Modern Physics has taken tremendous strides and has established a system based on scientific rules and the vast knowledge acquired by experimentation work done and being done by scientists all over the world, well documented, reported and discussed in seminars by eminent scientists as well as tested in the laboratories. The Vedic texts say whatever they say, in a few mantras. No record of any experimental support is available. Sāṅkhya treatise is of help upto a certain extent. Modern Physics is a highly tested science and its structure is laid upon well defined settings. We have, therefore, tried to locate points in Vedic Physics which, when plotted, appear to draw a line parallel to the modern science or otherwise as the case may be.

As no other corroboratory material was available, we were required to take recourse to a methodical approach in respect of the interpretation of Vedic text and terms. The age-old prescribed system is to avoid any conventional or traditional approach and to adhere exclusively to the etymological interpretation instead. This is the method established by ancient scholars and advised by Maharṣi Dayānanda as well. In Sanskrit this is known as the *yaugika* method. The word is interpreted through an analysis of its derivation in the light of the field of meaning prescribed for its root in conjunction with any prefixes/suffixes used as per the rules of grammar.

But Vedic terms are more than *yaugika*. They are ***yoga-rūḍha***. This distinction must be clearly understood before any consideration of main text, which follows, is attempted. Our dictionary has included both the terms *yaugika* and ***yoga-rūḍha***, but has failed to explain the latter one properly. The term *yaugika* has been stated to mean etymological which is fair enough. But ***yoga-rūḍha*** has been shown to denote a special as well as an etymological meaning. Such an explanation leaves a very wide field open to confuse the specific import of meaning the Vedic term presents. The etymological basis displays the characteristic constitution of the term. On the contrary ***yoga-rūḍha*** is a compound of two words, *yoga*+***rūḍha***. Its 'yoga' part is a clear indicator of etymological basis, but the latter ***rūḍha*** denotes the specific meaning which the term concerned has been derived to convey. It is not a special and etymological meaning a Vedic term denotes, but rather *a specific meaning covered under the range of its etymological derivation*. Hence a Vedic term denotes a specific meaning and not any meaning covered under its etymological construction. Such a specification may change with the context in Veda. That is why conventional meanings lead one astray while studying a Vedic text.

In order to avoid dry and lengthy explanations on every occasion the terms *yaugika* which is far more common, or etymological/ derivational or the like have been used. At all of such locations the respective term must be taken to denote the specific meaning pointed out. The use of term *yaugika* should be taken to convey the sense of ***yoga-rūḍha,*** a term less common in usage.

It has been held by the ancient ṛṣis that Vedic terms are ***yoga-rūḍha,*** and, therefore, the same must be interpreted in the light of their derivations and contexts. The Vedic language is confined to Vedic texts only. In subsequent literature if any term happens to appear beyond the rules of grammar, it is justified as *ārṣa* usage, *i.e.*, used by a '*ṛṣī*'. Such a usage is treated as an exception only in respect of the work of some '*ṛṣi*'. Otherwise Sanskrit language permits no liberty. Any off-track usage is an *apabhraṁśa* or a corrupted word fallen out of the said language.

Under the said ancient method of interpretation applicable to Vedic texts, the lexicons are not of much help. The lexicons in Sanskrit mostly state synonyms which are quite rich in conventional meanings that have come to be attached to them. Interpreting Vedic texts with their help may prove either misleading or going astray.

The capacity of a Sanskrit term to contain meaning, is enormous, and is revealed only when analysed etymologically. But even such an approach remains wanting in respect of numerous terms and expressions used in Veda. In this field only the great work of *Yāska* enlightens our path. This is two-fold, named *Nighaṇṭu* and *Nirukta* respectively. The first is a grouping of various Vedic terms under various heads. The second is virtually a torch that enlightens one's progress through Vedic texts. That is why the same has come to be called *Vedārtha-dīpakaḥ*, *i.e.*, the lamp that enlightens the meanings contained in Veda.

Truly speaking *Nirukta* was a well established system exclusively meant for the study of Vedic texts. Many books were written by a number of scholars. References of quite a few are available in the work of Yāska. Unfortunately they have all been lost, and the *Nirukta* by Yāska is the only work left to guide us. But it is a very competent master to teach one how to dive or swim in the ocean of Veda. Without recourse to *Nirukta* of Yāska no claim to the study of Veda is sustainable.

Besides all this, a few scholars have given some rules for the purpose of guidance in such an endeavour. They are quite useful and rewarding. The most important among them is the rule that all Vedic texts are liable to be interpreted in consonance with the laws of nature, as Veda does not convey anything contrary to or in contradiction of such laws. This set of rules is being reproduced below:

Rules for the guidance of Vedic interpretation:

1. The Veda states universal rules, none that may tend to create conflicts among the human society.
2. There should be no reference to any specific country or race therein.
3. Nothing should be in contradiction of the attributes and characteristics of the Supreme Being and/or the rules of Creation.
4. It should clarify that Veda is wholly a book of knowledge and science.

5. It should be complementary to all human requirements related to knowledge.
6. It must be in consonance with the rules of logic and proper argumentation.
7. It must not be the history of any specific person, race or country.
8. It may be helpful for a three-way interpretation, *i.e.*, physical, intellectual and spiritual.
9. It ought to be supportive of earlier precedents laid down by great scholars, and in line with the views held by *ṛṣis* and *munis* (seers and thinkers).
10. No order should be in conflict with any other order.

Thus far we have tried to point out that ancient *ṛṣis*, who are believed to be the seers of Vedic *mantras* and other scholars had emphasised the etymological interpretation of Vedic terms. Such a study must be in consonance with Yāska's works, namely *Nighaṇṭu* and *Nirukta,* both of them being necessary for the benefit of the study of Veda. And finally we have mentioned the beacon rule that any interpretation of Veda must not transgress the laws of Creation. Such is the stand of Indian authorities in this regard and we have tried to follow their directions.

But it would be interesting to note what a famous British lexicographer, who, according to his own admission, took upon himself the responsibility of such an assignment avowedly to uproot Veda and Indian wisdom contained in Sanskrit literature, has to say regarding the malleability of Sanskrit roots to give expression to the meaning ingrained in them. This author is Sir Monier Monier-Williams, M.A., K.C.I.E., Boden Professor of Sanskrit, Oxford University, who had authored the famous Sanskrit-English Dictionary, first published in 1872 and the second edition in 1899.

In his Preface to the New Edition in the second edition of his work mentioned above Sir Monier Monier-Williams, on page IX, says:

> "In explanation I must draw attention to the fact that I am only the second occupant of the Boden Chair, and that its Founder, Colonel Boden, stated most explicitly in his will

(dated August 15, 1811) that the special object of his munificent bequest was to promote the translation of the scriptures into Sanskrit so as to enable his countrymen to proceed in the conversion of the natives of India to the Christian Religion."

It may be added that Sir Monier Monier-Williams also authored an English-Sanskrit Dictionary.

Now we quote below what Sir Monier has to say regarding the Sanskrit roots in his Introduction to the same edition mentioned above. In the first section on page XIII he says:

"... a Sanskrit root is generally monosyllable, consisting of one or more consonants combined with a vowel, or sometimes of a single vowel only. This monosyllabic radical has not the same cast-iron rigidity of character as the Arabic tri-consonantal root before described. True, it has usually one fixed and unchangeable initial letter, but in its general character it may rather be compared to a malleable substance, capable of being beaten out or moulded into countless ever-variable forms, and often in such a way as to entail the loss of one or the other of the original radical letters; new forms being, as it were, beaten out of the primitive monosyllabic ore, and these forms again expanded by affixes and suffixes, and these again by other affixes and suffixes, while every so expanded form may be again augmented by prepositions and again by compositions with other words and again by compounds of compounds till an almost interminable chain of derivatives is evolved. And this peculiar expansibility arises partly from the circumstance that the vowel is recognized as an independent constituent of every Sanskrit radical, constituting a part of its very essence or even sometimes standing alone as itself the only root."

Sir Monier had acclaimed the help and support made available to him in this endeavour by the Secretary of State to the British Government and three Viceroys in India during his three sojourns regarding the execution of his assignment, more specifically the generous help by the Secy. of State to enable the copies of the work by Sir Monier made available at a cheaper price.

We would like to state that in spite of such a tendentious approach, we have accepted the said Sanskrit-English Dictionary by Sir Monier Monier-Williams for the purpose of choosing and selecting English equivalent for quite a few words used in Veda and other Sanskrit books, wherever we felt that they indicate the meaning of the concerned term in Sanskrit. But at places where either the said author has not followed the meaning or else he has attempted to fulfil his assignment by carrying out a mistranslation, we have taken recourse to the etymological explanation and explained the grounds for such a treatment on our part. In this regard the dictionary by Sir Monier has been referred to as *our dictionary* or *the dictionary*. But in this book we have preferred to use Indo-Romanic equivalents for sanskrit terms and quotes in Devanagari letters as per the manner printed on pages vii, viii, ix in the initial part.

Ever since the advent of Albert Einstein Modern Physics has stepped closer to Veda and the theories concluded by Stephen Hawking, and quite a few Nobel laureates on the subject prior to him, have further contributed towards the same direction. We have no intention of suggesting that the scientists mentioned above and those of their ilk are doing so deliberately. They may not even have seen any of the Veda, not to say of studying the same. Moreover the study of Veda is not easy like going through a book of fiction. One has to learn it the hard way. But the profound exercises undertaken by others to dissect Veda to help the cause of Christianity were proficient enough to gather the direction and/or the volume of contents related to scientific expressions contained therein, knowledge of which might have proved handy to researchers.

It is also true that the change of direction to the investigation of Physics provided by Einstein, and the fundamental questions the answers to which are being sought by Stephen Hawking do remind us of the questions that Vedic ṛṣis did ask and answered in their quest to know and realise the reality behind the appearance of this ever changing universe. The *Śvetāśvatara upaniṣad* begins with the questions which appear to be resounding through the quest of Dr. Hawking.

We believe in what Maharṣi Dayānanda declared, that is, Veda is the book of all true sciences. But we are given the right to examine it, and the test of such a belief must rest upon the hardest rock of truth. All exploration by modern science is also motivated with a sincere thrust for determining the truth. The truth never has a duplicate of its own. All expeditions for truth must converge on the common ground of reality unless, of course, they are led astray. It is, therefore, not surprising to see trends developing in the quest by modern science turning towards a point of convergence with Veda.

A study of Vedic Physics in the light of the commanding heights attained by Modern Physics, if succeeds, may counter the universal misconception regarding Veda being a scripture of some ancient religion, prepared and foisted upon the people by so-called foreign rulers with the object of perpetuating their rule here as well as of the so-called enlightened vested interests that have cropped up in this land, whose vision is blurred on grounds other than genuine and who are afraid that a Vedic revival might injure their interests. On the other hand the scientific community at large may feel inclined to tap this ancient source of concentrated knowledge called Veda, may give it the respect due at least to a book of true sciences, and come forward to help the human race benefit from its contents.

In such a study of Vedic Physics it must be remembered that Veda presents the knowledge in a highly concentrated form of a seed which contains within itself the capacity to be able to grow into an enormous tree provided properly nursed and worked upon. But each of the commanding height of Vedic knowledge dips and expands into a vast landscape integrated with the vast expansion. There is no intention to match Vedic Physics with modern science. The attempt is motivated by the desire to show that Vedic Physics is the seed of true knowledge of high variety and potential and deserves respectful attention full of purpose. It has grain that ought to be gainfully harvested.

We have, therefore, selected the elements of creation enumerated by Sāṅkhya, which present a wholesome structure of Vedic Physics. But our study is in the light of Modern Physics, hence the elements related to the biological development of matter, enlisted therein, have been given up. These in all total eleven which, when taken out, leave only fourteen for our study. They are Prakṛti, Mahat, Ahaṅkāra, five Tanmātras, five Mahābhūtas (or Sthūlabhūtas as called by Sāṅkhya), and Puruṣa. Besides these a few subjects have also been included from Veda itself. These are Ṛta, Satya, Kāla, Pralaya, Mahat Sphoṭa, Yajña, and Vedic Calendar. A chapter regarding the birth of universe as well as one on Vedic Model and an Epilogue have also been added by way of a recapitulation and conclusion.

We may as well add a few words by way of an introduction and then proceed with the subject directly. All knowledge is there in the Creation. It comes only through learning and observation/meditation. When somebody stumbles upon or realises any aspect of such knowledge his ego makes him feel that it is his knowledge. The fact is that we roam about throughout our life thinking that we are the possessor of intelligence which is part of our personal existence and as such whatever we pick up with the help of our mind is our property or earning. Like the Church and the rest of the world who were blind to the fact of earth being heliocentric when Galileo, based on his own observations realised the reality and declared the same, he was taken to task and forced to apologise for pointing out the truth which was considered not only untruth but even blasphemy, we too, fail to realise that knowledge is universal.

The realisation is the revelation of such knowledge. That is the only way. Such a revelation may come through intuition or through deep meditation in the state of samādhi by the way of yoga. The word *yoga* literally means union and scientifically denotes union with the manifestation or its Efficient cause. Even the speech comes by learning and is not a part of what is called instinct. Veda is the knowledge revealed to or realised by the four ṛṣis in their samādhi, *i.e.*, deep meditation.

Let us take an example. Modern science has established that planetary orbits of our solar system are elliptical, and so are the bodies of these planets. But Veda says that the orbital motion being the result of centripetal and centrifugal forces working upon a body in motion, all orbits are elliptical and their bodies too. This is a universal phenomenon.

But respective calculations may be done on circular basis. The Vedic noun for an ellipse is *triṇābhi* or triple-centred. A circle having one centre is called a Cakra. By using the term **triṇābhi cakram** in the R.V. 1.164.2 Veda sums up what we have explained above;, *i.e.*, all knowledge is in the Creation which we learn by observation and realisation.

The period of scientific awakening in Europe begins from the time of Galileo (1564-1642). At that time the Church was declaring such claims of findings which did not conform to the Bible as blasphemous. Galileo too had to suffer and apologise. In 1615 Sir Thomas Row had obtained the royal 'firman' from Emperor Jahangir and the East India Company soon came to be pretty well entrenched in certain parts of India. Newton was born in 1642 the year Galileo died, and breathed his last in 1727 by which time the era of Aurangzeb was long over and the Moghul Empire had become a hot bed of conspiracies. In another 50 years or so the British Company was to be a very prominent political and military power after the victory of Plassey in Bengal. Soon it was going to have complete sway over India.

Early in the 19th century Lord Macaulay enforced his plan for the cultural enslavement of India. What attracted attention most was his enforcement of English language as the compulsory medium of instruction in education. His intention was to subjugate India to the British culture. The second blade to this scissors designed and employed to cut off India from its immensely rich ancient heritage, was far more diabolical and cunning. It worked almost unnoticed. It was the setting of the curriculum for education in the way that Science and Sanskrit were made optional subjects. This cut still deeper. Those students taking Science were debarred from opting for Sanskrit as well and vice versa. For those taking Science, lots of avenues were open. Those opting for Sanskrit were cut off from developing any modern scientific perspective. They had hardly any avenue save a teacher's job. English and Science both had their source of inspiration in England. The whole generation was being attempted to be anglicised.

Simultaneously search was on for the collection of ancient books of reverence and knowledge in India to understand the cultural psyche of the majority population. The same were sent to England for a thorough study with the object of devising an antidote or to discredit them as primitive views of a primitive people having no knowledge of science and technology. The European states which had also succeeded in establishing colonial rule in parts of India prior to them, like the Portugese, French and Dutch, who, in years to come, were either driven out or confined to small territorial existences by the British in Indian landscape, were also busy in this regard. But the British by the end of the 18th century had succeeded in eliminating or confining their European rivals to ineffective existence, and had become number one political power. To justify their invasion and establishment of their empire in India the British in a concerted manner started the myth of Aryans being the first invaders of this subcontinent for colonising the same. Even the mission of Sir Monier Monier-Williams quoted above is a clear indication of this vile campaign undertaken to malign Veda and its antecedents.

Under the circumstances, Dayānanda played an effective part in rewinning the attention of not only the people of India but also of the intellectuals all over the world for Veda and restoring the same to their pristine position among the masses here. He declared that he had obtained a copy of Veda from Germany. Unfortunately, the virus of British campaign and their legacy left behind in Independent India, have persisted for over five decades and are vastly responsible for the state of inertia in respect of Vedic research, and still continues to be an effective hurdle. There is no systematically organised effort, supported by substantial resources provided by the Indian Government or otherwise to delve deep into Veda and to explore them for scientific advancement.

It must be said that a few Western scholars, who came in contact with Veda have been quite eloquent in their eulogy for them. Even Max Muller, whose poverty forced him to accept a job at the Oxford University to work in accordance with Macaulay's design, later on tried to make amends and even admitted his earlier role unworthy as a scholar. One Louis Jacolliot, who happened to be the Chief Justice in the French colony of Chandra Nagar in India in the 19th century, wrote a book under the title *'La Bibledans and Inde'* which appeared originally in French in 1868; its English translation under the name *'Bible in India'* appeared in 1869. In it, he wrote that Āryāvarta was a storehouse of knowledge and goodness and all sciences and schools of thought had spread out from this very country. In it he even prayed 'O God ! Kindly make our country as prosperous and advanced as was Āryāvarta in earlier times'. The magic of Veda is such that even Sir Monier Monier-Williams, in spite of being committed to a biased assignment, praises the Sanskrit roots and the structure of the language in placing the system at the highest pedestal as is evident in the quote from his Introduction to his dictionary, mentioned earlier above.

The modern, scientifically advanced world is more and more realising the superiority of therapeutic qualities of the Indian system of medicine, the Āyurveda, as its medicines generate no side-effects. It has its source in Veda. Indian music is esteemed to be of the highest order. It has its origin in Gandharva Veda which originated from Sāma Veda. The psalms, pronounced as sāmz, a book of the Bible, composed of 150 songs, hymns, and prayers is still a pointer and reminder of its original source, by still carrying its name as Sāma under a misspelt form. It further indicates that hymns and prayers to God or the Supreme Being were called 'sāma' or "psalm" before the arrival of Christ. This speaks of some sort of Vedic association. The gospel in its infancy, might have been considered a school of spiritual reforms associated with the teachings of Veda in its infancy and later on came to be recognised as a separate religion after his name when the Church asserted its origin as a separate religious entity.

The word *Vimāna* (aircraft) comes directly from Veda. The word means a carrier built upon bird-engineering. A book entitled *Bṛhadvimānaśāstra i.e.*, greater aircraft engineering, has been found. It defines an aircraft as a means of travel within the country, from one country to another country, and from one planet to another planet, the last at present falls under the term space-travel. The book has indications about inter-galactic travels. The modern earth satellites were also known. One anecdote refers to Ṛṣi Viśvāmitra to have established a *tri-śaṅku* what is known today as a stationary earth satellite. The name *tri-śaṅku* reveals the technique of such a launch, that is why the anecdote cannot be dismissed as a fancy of imagination. *Tri* is three and *Śaṅku* is a conicular body. Together the compound word means a three-stage conicular body, which has now become so well established due to modern science. The most pertinent part is the use of 'three' and the conicular shape of modern rockets. The scientists have determined that to overcome the pull of earth's gravity, direct firing from the surface fails to provide necessary lift. But a three-stage firing provides a conicular body the required thrust. Obviously tri-śaṅku is not an imaginary concept. The space launching rockets of today may well be named *tri-śaṅku.* These illustrations are to emphasise the etymological approach to gain the scientific meaning from Veda.

Returning to the subject of Physics and with due apologies to the giant physicists of modern scientific world, it is most humbly submitted that in spite of their landmark contributions to the knowledge of physics, the subject is neither being treated nor being taught or studied as the Science of Physics. It is more being presented as what may be called the history of the development of physics. Even the vast and creditable experimentation work conducted by worthy physicists is reported like the details of some excavation expedition carried on to locate some hidden mystery. It is not a presentation in a cause and effect sequence which is the fundamental foundation of the entire process of Creation. The names and years may be important factors in any narration or reporting of history, not a science.

A science is, on the contrary, a precise and systematically graded presentation of the subject stepwise, each step leading to the succeeding one revealing the whole system. 'Sāṅkhya' treatise presents the Vedic science of Creation (read Physics) in a precise sequence of cause and effect. Physics is the science presented in cause and effect sequence since the beginning and that is how it ought to be presented, and has been presented as such by Sāṅkhya. The names of scientists may appear to pay homage in the course of details as the seers of the particular truths, because all science is the knowledge of truth. Galileo did not make the earth heliocentric. It was in such a state since its origin. But people, due to a misplaced belief, considered it to be the centre of the universe. Faith and belief ought not to precede science; they must follow it.That is the Vedic approach, and that is the only correct and scientific approach.

To clarify the thrust of above observation, the sūtra 1/61 of Sāṅkhya is being reproduced below followed by its English translation. This sūtra enumerates the elements of the *sṛṣṭi-vidyā* (Science of Creation, i.e., Physics). It says: *sattva-rajas-tamasāṁ sāmyāvasthā prakṛtiḥ, prakṛtermahān, mahatohaṅkāraḥ, ahaṅkārāt pañca-tanmātrāṇi (ubhayamindriyam), tanmātrebhyaḥ sthūlabhūtāni, puruṣaḥ, itipañcaviṁśatirgaṇaḥ.'*

Translating in English and substituting the terms of the Modern Physics the above sūtra would read something like the following:

"In existence, motion and transformation what remains conserved is energy; from the energy emerges the primordial fireball; from the said fireball emerges the multiplicity of identity; from the mulitplicity of identity emerge five grades of quantum levels; (double number of senses and organs of action); from quantum levels emerge the five mega-causatives of relativity and Puruṣa." Sāṅkhya presents Vedic Physics within this brevity. These are the commanding heights which cover the nuclear physics, the particle physics, the quantuṃ physics, the physics of relativity, the astrophysics and more as we shall see as we proceed. The bracketed portion deals with the biological creation and hence has been left out. The Puruṣa is also a double, one related to the cosmos, and the other to the biological world. Hence the latter has also been left out, save the places where the continuity of Veda mantras, quoted and translated to explain the details of some subject, include any reference to the second puruṣa. There the same has also been translated for the benefit of the reader and to avoid any interregnum in the quotation.

It may further be noted that Energy has no cause of its existence. It eternally exists, and conserves itself in any or all of the three states of its being. From Energy commences the cause and effect sequence as far as the causatives related to the field of relativity. These form the workshop of creation wherein matter gets transformed from Energy, the material cause of the universe, moving under the stimulus of Puruṣa which again is causeless, in eternal existence and is the instrumental cause of Creation. An interaction between Puruṣa and Prakṛti (Energy) results in the above workshop coming up and the universe being churned out of the same.

Puruṣa is the most important factor that causes the emergence of cosmos. We have, therefore, reproduced the entire *Puruṣa sūkta* from Y.V. and discussed the same somewhat in detail in an exclusive chapter. According to Maharṣi Dayānanda, the cosmic development is a planned execution. This aspect will be discussed in the said chapter. Two other sūktas have been quoted in full from R.V. in order to show what Veda says regarding the substance within the primordial ball of fire as also the sequence of the radiation after the so-called Big Bang, transforming itself to the stage of the first generation stars. The evidence revealed from the depth we have been able to dive, to say the least, is self-explanatory and demands greater research by competent scholars. But apart from the three sūktas mentioned above various mantras have been quoted in parts or full and discussed at length in respective contexts. Apart from Veda, quotes have been made from Sāṅkhya and references have been made to Vaiśeṣika as well.

As pointed out earlier, quite a few subjects have also been included to present a comprehensive view of the subject. Usually it has been preferred to use the terms being used by the modern physicists. But at certain places comparisons have also been made to illustrate the superiority of Vedic terms being more comprehensively expressive or specific. Exceptions have also been made by using the English translation of a Vedic term instead of the term in vogue at present. This has been done to do justice to the sense conveyed by Vedic term. One such example is the use of the noun *dissolution* for Vedic term *Pralaya.* The term in vogue with the modern physicists is recollapse. The Vedic term, barring the prefix *pra*, communicates the sense of melting. It implies the idea of conservation as, taking an example, the amount of water frozen into ice or as liquid remains the same. This aspect is strongly emphasised by Vedic term *'laya'* (dissolution). The prefix *pra* carries the scientific meaning contained within the term still further. It adds the sense of onwards communicating that dissolution is not the end of the process of Creation, but a continuous spiralling motion which moves on to the re-emergence of the universe in good time and so on. This having been explained at the proper place the English noun dissolution has been used as a synonym for *'pralaya'*.

The term *Veda* has been used normally as the same is current but the anglicised form Sanskrit has been preferred to the correct form Saṁskṛta, as the former has come to stay that way in English. All other Sanskrit or Vedic terms mentioned have been marked by single inverted commas followed by their meanings or explanations immediately thereafter. The intention is to provide an opportunity for the Sanskrit knowers to assess their interpretations as well as the conclusions drawn from them, if any. Quotations from Veda, Upaniṣads, Sāṅkḥya, etc., have been preferred in bolder type. Other quotations have been marked.

This presentation is in no way final. We stand convinced that Vedas are more or less an immense source of knowledge to enlighten humanity in numerous ways. We are aware of our handicaps. We feel that a person properly educated in the ancient system of Vedic education as well as in modern science with the benefit of experimentation, may be able to approach Veda with a far more penetrating insight. We feel that many sūktas in Veda may reveal far greater details that may help throw light in areas still under research. The modern science may verify such indications through experimentation or otherwise and the humanity may stand benefitted that far and that soon. The onus rests much more upon such Indian scientists who happen to be acquainted with the grammatical structure of the Sanskrit language. They may more efficiently adopt the etymological approach, and their *ūhā,* visionary conjecture, as a scientist may help them unlock such Vedic terms and gain the knowledge contained therein.

— 1 —

PRAKṚTI – THE MATERIAL CAUSE

What is this Prakṛti which the Sāṅkhya enumerates as the first element for Creation? The synonym that is commonly used in English language for this Sanskrit word, rather a Vedic noun, is nature. But basically nature is the instinct or tendency, individual or collective, as the case may be. In the context of discussing the elements of Creation the English synonym mentioned above does not fit. Creation is a scientific subject, and the Vedic noun Prakṛti ought to be analysed to determine what it connotes. To facilitate matters, Sāṅkhya has virtually defined it in unambiguous terms, which also goes to state a well established law of Modern Physics. Let this angle be considered first, as that would also be of help in analysing various derivations as well.

The Sāṅkhya says **'sattva-rajas-tamasāṁ sāmyāvasthā prakṛtiḥ' 1/61**. Translated it means Prakṛiti is that which remains in the state of conservation in existence, motion and transformation/condensation. Let each term be expounded for the purpose of clarity and better understanding.

Sattva: The *tva* suffix renders the word an abstract noun of *sat*, the latter meaning present or eternally present. Presence presupposes existence. Our Dictionary states *sattva* to mean being, existence, entity, reality. All the four meanings given are akin, but existence has been used in this translation to convey the sense.

Rajas: The rule of grammar which adds suffix '*as*' to the root *rañj*, to attach, makes the *n* invisible, and the noun *rajas* comes into being. One of the meanings of the said root is to move. Hence this derivation connotes motion.

Tamas: This noun is also derived with the same suffix '*as*' as in the previous para from the root *tam*, and means inclination. The lexicons explain that the inclination is towards transfer, transformation and/or condensation. The *ām* in the quote denotes the sense of being related to.

Sāmyāvasthā: This is a two words compound, *sāmya+avasthā*. *Sāmya* is the abstract noun of *sama*, literally meaning sameness or equality, *avasthā* is the state of being. Thus the compound means that the state of being remains the same or equal or conserved.

The two compound words define Prakṛti as that which under any of its three states, *i.e.*, the state of existence, or of motion, or that of transfer/ transformation or condensation to some other form remains the same. It does not undergo any loss or gain. In other words Prakṛti cannot be created, nor can it by any means be annihilated. Under any state of being it remains just that much as it was prior to the said state.

What is this Prakṛti then? The Vedic seers say that it is the *Upādāna kāraṇa* or the material cause of Universe or Creation. The said material cause can be identified in any of its three states so named. First it exists, is eternally present. Secondly it may be in a state of motion. And thirdly it is capable of transferring or transforming itself into other forms. But throughout such a behaviour or activity its total entity remains the same. It neither loses even an iota nor adds any to itself.

The three states of being are called *guṇa* of the said Prakṛti. As per the dictionary mentioned above the word *guṇa* means a quality, peculiarity, attribute or property. So Prakṛti has three attributes, namely, eternal existence or presence, motion and transformation or condensation. It has no other attribute. They come into play as Prakṛti causes the material universe to emerge and function. But Veda says that Prakṛti is *jaḍa* or inert. An inert entity can certainly exist, but it will remain totally incapable to undertake any activity whatsoever on its own. If an inert object displays activity, the cause must be sought elsewhere. This brings one to the conclusion that the other two *guṇas* or attributes come into play only when they are impelled by a factor other than Prakṛti itself. In other words Prakṛti remains inert and behaves as such even when under the influence of its second and third *guṇa* (attribute). The activities of motion, etc., are caused to be undertaken by it under the influence of some other entity or being. So far as Prakṛti is concerned, under the influence of any of the three attributes it stays inert and continues to behave that way. It may also be concluded that the attributes *rajas* and *tamas* stay dormant within *sattva*, and appear only when Prakṛti is caused to materialise and further to the extent so caused. This cause has been most forcefully advocated by Veda as the source of all sciences and is called *Puruṣa* enumerated by Sāṅkhya as last, which may get mentioned in between but will be considered and discussed at some length at a later stage.

Modern Physics, too, recognizes the material cause of universe and names it 'energy'. The energy exists eternally. It was, it is and it shall eternally continue to be. It is not born, nor does it die. It can neither be created nor annihilated. No gain or loss accrues to it. In other words, it conserves itself.

There is a law of conservation. It runs as follows:

> Energy cannot be created or destroyed. It may be transformed from one form to another, or transferred from one object to another, but the total amount of energy is constant or conserved.

Compare this law as stated above with what Sāṅkhya says. The first sentence, *i.e.*, "energy cannot be created or destroyed" just states the first attribute stated by the Sāṅkhya, that is, sattva (existence). The part portion of the second sentence reading it may be transformed from one form to another is the third attribute named *tamas* by Sāṅkhya, and finally the mid-portion, *i.e.*, or transferred from one object to another, is what the second attribute called *rajas* is conveying. Finally the last part of the above quoted law reading that the total amount of energy is constant or conserved is just stating the meaning of *sāmyāvasthā* (the state of being the same or equal). The only difference is that Sāṅkhya has stated the three attributes in respect of what has been named by it as Prakṛti, while Modern Physics is asserting them in respect of energy. Hence, before we move further, let us try to clarify this situation.

According to Modern Physics all matter is energy. The Einsteinian formula $E = MC^2$ provides the key to calculate how much energy is locked up in any particular form of matter. As such the entire matter of the universe is but a condensation of energy. In other words, energy is the very cause pointed out above which according to Veda is Prakṛti *i.e.* the '*upādāna kāraṇa*' (material cause) of creation. Here again the two nouns, one from Modern Physics and the other from Veda, converge on the same activity, *i.e.*, creation of matter.

As mentioned above, Veda declares Prakṛti to be *jaḍa* (inert). The Vedic position is that under the influence of any of its three attributes, Prakṛti remains inert. According to the lexicons an object that is jaḍa does not know what is desirable or undesirable, nor does it feel any comfort or pain. It moves under the control of somebody else.

Modern Physics declares energy to be inertial. Such a declaration was first of all made by Galileo. Then came Newton. He gave his laws of motion, the first of which is called the law of inertia and runs as follows:

> Every body preserves its state of rest, or of uniform motion in a right line unless it is compelled to change that state by forces impressed thereon.
>
> Subsequently this law has been further modernised:
> An object remains at rest or in motion with a constant velocity unless acted upon by an unbalanced force.

All of this goes to show that whatever is being described by Veda about Prakṛti by pointing out the latter's three *guṇas* (attributes), or by mentioning it as the *upādāna kāraṇa* (material cause) of creation, or by stating its nature to be '*jaḍa*' (inert) is being repeated by Modern Physics in relation to energy.

Another noteworthy stand of Modern Physics is regarding the state of being of energy. When energy is, so to say, at rest, it is in a state of equilibrium. It is sort of a motionless state balanced all around. In this state of being energy remains totally at rest. But it most certainly exists. That is why Sāṅkhya has named it *sattva* (existence). The energy in motion is named kinetic energy (K.E.), the adjective kinetic being self-explanatory. In Vedic terminology this is the state of *rajas* as has already been explained earlier. Lastly, the physicists declare that all matter in whatever form is energy condensed within the appearance of that form.

Each and every object is a storehouse of energy, and in any position the energy forming the same being its potential is called potential energy (P.E.). As discussed earlier this is the state of the third *guṇa* of Prakṛti named *tamas* by Sāṅkhya.

This discussion, therefore, appears to lead to only one conclusion that whatever Veda is stating about Prakṛti, is being confirmed or corroborated by Modern Physics in respect of energy. In other words what has been named as Prakṛti by Veda has been recognized as energy by the physicists of today. The two of them are synonymous, the former coming from the Vedic text, while the latter is in use with Modern Physics, but both denote, the same element in their respective languages.

It may further be of interest to note that Veda intends to emphasize some additional characteristic of the said element by employing the noun Prakṛti which is no less fundamental and basic to it than the import of the meaning of energy. In the Vedic literature another noun has also been used which is more akin to the English term energy which has been derived from a Greek origin meaning activity. This is *Śakti* which is an abstract noun from the root *śak* denoting power. Energy too is a synonym of power. That is why *Śakti* means capability or strength to do or execute the desired work or activity. At numerous places this word has been used to convey the sense of energy. And yet Veda preferred to name the *upādāna kāraṇa* (material cause) of the universe as Prakṛti. Let us, therefore, analyse this word itself.

Prakṛti is *pra+kṛti*. The *pra* is a prefix meaning, forward, in front, on, forth, and *kṛti* means, act of doing, making, performing, manufacturing, composing etc. As the material cause of universe the *kṛti* part in the said noun is bound to mean the act or acts of creation of matter. But the pra prefix unmistakably shows the direction of creation, which is forward. Hence Prakṛti in a continuous forward or onward motion creates, maintains and destroys to create anew or in various forms the material bodies or forms. The Vedic term, therefore, clearly states the direction of the performance of energy as well, which the term energy does not convey. It is clear that Prakṛti of Veda or energy of Modern Physics, creates, maintains and destroys the forms of the matter through transformation in a state of forward motion all the time.

This uniqueness of its characteristic activity is also so very well expressed by this Vedic term.

Secondly, in Sanskrit grammar all the roots are treated as unmanifested. A root gets manifested in the specific form of a word when a suffix gets added to it. The word obtained is the creation of that root. The state of unmanifested root is called *Prakṛti* which awaits the addition of a *pratyaya* (suffix) for the purpose of manifestation in the form of a word. For the cosmic creation in the form of material system in its multiple variety, the source of all creation has as well been named *Prakṛti* by the Veda and the rest of elements enumerated by Sāṅkhya, except the last one, are the influencing factors acting like suffixes, the creation being called the *kāvya* (poetry or verse) of the Creator by Vedic seers:

devasya paśya kāvyam na mamāra na jīryati (A. V. 10.8.32)

Look at the poetry, superb composition, of the Creator; it neither dies nor does it get worn out. Such a statement points to the smooth order prevailing in universe. This sense is communicated by the noun *kṛti*.

Thirdly, by defining *Prakṛti* by mentioning its three attributes Sāṅkhya in a nutshell is speaking volumes about Prakṛti. The very first, sattva virtually covers the entire range of existence. Whatever is possible, nay, even conceivable, is achievable under *sattva*, which denotes the capacity. The *sattva* is the fundamental base for all kinds of activity and the very motion related to the same. The inertia is not in conflict with its capacity to take to motion or transformation. Like the mathematical zero being capable of giving out the entire lot of numerals from 1 to 9 and finally rolling them all in itself, so is the *sattva* in respect of creation. Even the Sanskrit term *satya* for truth has to have the root support of *sattva* to be *satya*.

Fourthly, the second attribute *rajas* denotes that the entire capacity of Prakṛti to manifest itself in innumerable forms of waves and matter is based upon its attribute *rajas*, that is, motion. The emergence, maintenance and destruction or the final dissolution, as stated by Veda, are all the results of motion. *Rajas* is derived from the root *rañj*, but is specified to denote motion. Moreover the root *rañj* also denotes colour. As such the term *rajas* further implies a motion containing colours. The modern spectrograph supports Veda in this regard.

Fifthly, the third attribute *tamas* meaning inclination for transformation conveys the sense that motion is inclined to achieve the purpose of the direction. The motions of Prakṛiti become purposeful even though by itself Prakṛti is incapable of possessing any such capability. The Prakṛti, according to Veda, is and remains the material of creation quite incapable of understanding anything being achieved through its activity.

Another Sanskrit term related to creation, yet traditionally much maligned and misunderstood is *māyā*. It has been and continues to be a subject of debates and discourses. It essentially depicts the physical aspect of creation. The term *māyā* clarifies the scientific meaning contained in the attribute *tamas*. It is derived from the root *mā* to measure, mete out, mark off, etc.. Essentially it denotes some form or assumption of form as the same is an object of measure. Measure communicates space-time boundary.

Vedic Physics describes Prakṛti as the material cause of creation enjoying eternal existence and capable of conserving itself in any of the three states of its being: existence, motion and transformation to any form of matter. Modern Physics knows it as energy. The Prakṛti in motion is named *ṛta* and its transformed or condensed state is called *satya*. In modern science energy in motion is known as K.E. and the condensed form is known to possess P.E. But the Vedic *ṛta* covers greater ground in performing the work of Creation *vis a vis* K.E. This aspect will be presented in detail in an exclusive chapter on *ṛta* latter. *Satya* will also be presented similarly.

According to Sāṅkhya and Veda before the initiation of the process of creation Prakṛti as the material cause was just in a motionless state of existence called *sattva*. That state of existence did not conform to the Einteinian formula $E=MC^2$, because there was neither M nor C. Only heat existed in a state below absolute zero temperature far beyond any conceivable measure. In that state in the manner to be described later, with the temperature starting to rise from the state under absolute zero, the second attribute *rajas* appeared putting it to motion. This second state is known as *ṛta* in Vedic context and K.E. in the modern science. The motion caused the appearance of magnetism along. Subsequently at appropriate temperatures the *ṛta* took to manifestation and following a bang the process of creation accelerated.

It may further be submitted that in the scientific context of its use the term *Prakṛti* is considerably more expressive than Energy, a trait which we may note more than once as we proceed further.

— 2 —

MAHAT

Sāṅkhya says, *prakṛitermahān*. Prakṛti caused *Mahat*. *Mahān* is a derivation of the seed word *mahat*. That was the very first effect, the initial job done by Prakṛti in initiating the process of creation. Modern science says that a work is done by the transfer of energy. *Sāṅkhya* later at 1/71 says that the *ādyakāryam*, the initial work, done by Prakṛti, *mahadākhyam* is called *mahat*. In other words Prakṛti (read energy) transferred (transformed) itself to effect *mahat* which was the first work done. This is another point where Modern Physics appears to be adopting Sāṅkhya or Vedic expression as the latter has already stated such a transfer or transformation as the work performed.

What is this *Mahat*, the initial work ? Till about a few decades back modern physicists began the subject of the development of matter from the state of hydrogen atom. They had unveiled the process of atomic growth from hydrogen onwards within the intensely hot oven that got created under gravitational force inside a body of hydrogen matter. But they remained clueless as to how the hydrogen atom came into being. The then conceived hydrogen body has now been substituted by a primordial fireball, a phenomenon so aptly described in Veda.

The noun *Mahat* is derived from the root *mah* which as per our dictionary means to arouse, to excite, to exalt. As *to arouse* denotes, to stir, to act, and *Prakṛti* being inertial, under its first attribute of existence in a state of rest, it could not do anything on its own. But *Sāṅkhya* says it caused the birth of *Mahat*. Naturally it means to communicate that *Prakṛti* was made or caused to be stirred into action, that is, take to *rajas* or motion, its second attribute. How ? By way of a *kṣobha* created, *i.e.*, stimulus caused. The source of such a stimulus came from *Puruṣa*, which is the instrumental or efficient cause. This would be discussed and quoted at length from Veda in a later chapter.

The stimulus resulted in the generation of a sort of pulsation or wave-like action. As there was no matter at that stage of existence, the stimulus came to be caused through a rise in the temperature that is inherent in Prakṛti even in the state of rest. According to the modern science, it is the state of being at or under absolute zero degree K. The temperature rise effected generation of motion. At this stage the first step of quantum physics appeared. A quantum of *Prakṛti* became operational and the stimulus started expressing itself in the multiplicity of the quantum unit both on the side of the temperature as well as the wave motion.

In the vast expanse of *Prakṛti* at rest, the stimulus caused a change in the part so affected. In Vedic terminology, it was the appearance of *Hiraṇyagarbha*, the one having *agni* germinating in its interior. The term *agni* denotes fire, temperature, radiation and all that goes with these. The term *Hiraṇyagarbha* is a compound consisting of two words: *hiraṇya* and *garbha* That which holds *hiraṇya* in its *garbha* is *Hiraṇyagarbhaḥ*. According to *Śatapatha* (4-3-4-21) *hiraṇya* is light or temperature while *garbha* means a womb or interior. Hence that which holds light or the source of light germinating in its interior is *Hiraṇyagarbhaḥ*.

The Veda says:

hiraṇyagarbhaḥ samavartatāgre bhūtasya jātaḥ patireka āsīt |
sa dādhāra pṛthivīṁ dyāmutemāṁ kasmai devāya haviṣā vidhema |

(R.V. 10.121.1)

Hiraṇyagarbha existed before everything else. It was the sole master of whatever was to be materialised. It held the earth as well as the entire illuminated bodies within itself. For that illuminated one called **ka** we provide to donate our very self and all that goes with it.

Veda states that this universe gets created conforming to a design or plan. The plan is conceived by the totality of wisdom and dynamism named Puruṣa. The creative system has not come into effect through a freak combination of circumstances. The Supreme Intelligence and

Wisdom is instrumental both in its plan designing and execution. It is not a one-time event, but keeps on repeating on an eternal basis in a cyclic occurrence.

Do we not realise that the universe is working as an automatically functioning system? All of its queerness, when gets explained on account of research studies, leads us to the subtle aspect of some complementary system or part thereof. There is nothing freakish about any of its motions, actions and/or reactions. Whenever the combinations happen to occur, result is expected. It would be too egoistic to presume on our part that we with the help of our intelligence and its persistent organised application uncover the queer mysteries of an unintelligent system known as universe, while also upholding the fundamental scientific axiom that something cannot emerge from nothing. If the universe is not based upon an intelligently functioning order, wherefrom did we receive our intelligence? The superb and orderly organisation gone into the cosmos got ingrained in it at its fetus state. And, that fetus state was Vedic *Hiraṇyagarbha* wherein the latter term *garbha*, literally meaning a womb, speaks volumes. Literally translated it means one bearing the fetus of light in its womb or interior.

No doubt the modern quantum physicists are right that Big Bang cannot be the beginning of the universe. They do not contest the occurrence of the event. But they maintain that the quantum nature of energy working forces the conclusion that the inception must have taken place earlier at a string-like level which ultimately grew into a huge size at the time of Big Bang. But it is Veda which puts it graphically and resolves the two schools of Modern Physics. Metaphorically speaking the seed or semen or, to be more specific, the spermatozoon of light rose to the imaginary ovary of *Prakṛti* when *Puruṣa* willed a rise in the inherent absolute zero temperature, effecting a conception which germinated in as the zygote of light. The temperature kept on rising and the zygote kept on growing till it emerged with a Big Bang.

As the modern science appears to be skeptical regarding the existence of a Supreme Being or *Puruṣa* on account of the commonly held belief regarding God, it may better be made most emphatically clear that Veda does not advocate or state even the ghost of an idea of such a being as is generally or conventionally understood. The Vedic Surpeme Being *Puruṣa,* known by any of its Vedic names, which are many, is a totally scientific and mathematical concept of existence, and need not be confused with any concept wavering in the slightest manner from scientific understanding or analysis. According to Veda science, even Physics, is not limited or restricted to matter exclusively. Energy itself in its state of pure existence at rest is no matter at all. Science does not, and must not, hesitate, to understand matter in its non-material existence. The Vedic *puruṣa* by its very nature, cannot even think of violating any rules of creation. Actually it is instrumental in devising all such rules and laws, and they are all eternal, none requiring even the slightest modification or amendment at any stage or time. Moreover *Puruṣa* even though instrumental in creating the cosmic system, does not interfere in its running, as the system so created is perfect in all respects and under any circumstances its perfection takes care of it under the laws of creation. This aspect has been elaborated upon in two separate chapters which appear later. The chapter on *Puruṣa* presents what Veda has to say about it. The other one on Supreme Master of Sciences elaborates in greater details what has been summarily observed above in this para. With this clarification we return to *Hiraṇyagarbha.*

As mentioned earlier, according to Veda, the creative process gets initiated and runs according to a plan which had been designed by the Supreme Being. This plan is eternal as *Puruṣa* in each cycle of creation initiates the process in accordance with it which finally culminates in the development of the entire universe. During the period when universe is no more on account of its *pralaya* or dissolution, the plan remains in its state of causation with *Puruṣa.* Veda states: **'yathāpūrvam akalpayat'** (R.V. 10.190.3), i.e., the cosmos was created as per its precedent.

Prakṛti is required to be transformed into almost innumerable forms for the creation of the cosmic system. This process gets initiated at the Hiraṇyagarbha stage with the formation of subatomic particles. As the interior of the Hiraṇyagarbha begins to develop into a furnace energy under the influence of rising heat and temperature goes on getting more and more excited and on the run. The leptons get manifested and with the intensity of temperature going very high, protons and neutrons also appear. In the whirl their actions, reactions and interactions get determined causing more and more particles and their anti-particles to materialise. Such matter and their anti-matter particles also get collided resulting in the total annihilation of both. On the other hand collision between a proton and a neutron effect their fusion in the form of deuteron. These actions, reactions and interactions form the basis of the development of atomic chemistry but that takes place after the Hiraṇyagrabha bursts forth in a big bang, and radiates its interior all over, initiating the next stage of creation. All of this would be described at some length in the words of Veda in chapter 8. Here what is intended to point out is that the actions, reactions, interactions that came to be born at this stage, were something like, if we may borrow such a term, instinctive behavioral development within the formation of a fetus in a womb inside a mother's body. In this case the mother was Prakṛti the interior of which was bearing the fetus of a coming universe.

The manifestation of such 'instinctive' behavioral pattern in the particles formed, which would mature at the later stage of atomic and molecular formations, is an indication of the intelligence input getting ingrained in the process of creation. That is why *Sāṅkhya* term *Mahat* is also interpreted as the *buddhi-tattva* or the element of intelligence. The sūtra 1/71 of the treatise part mentioned above, in full reads: **"mahadākhyamādyakāryam tanmanaḥ."** It means that the initial work of *Prakṛti* is called *mahat* which is based on an intelligent process. Quite a few students of Sāṅkhya get confused when they find the noun *mahat* to denote the first transformation of Prakṛti, and also as intelligence. In the above sūtra what Sāṅkhya is saying is that the transformation was got done on an intelligent basis. Intelligence is no material product; it reveals wisdom, and in the present case the divine wisdom of Puruṣa. The term *Mahat* further reveals that the product of the transformation was huge.

Veda names this first output by Prakṛti by way of transformation as *ka*. We have already quoted the first mantra of the Ṛgveda 10/121 above. The *devatā* or the subject matter of this sūkta referred to here is *ka*.

The term *ka* denotes so many things but here it communicates light including its splendour and heat. This has also been clarified earlier in the interpretation of *'Hiraṇyagarbha'* as a womb bearing the fetus of light. The Vedic term is all-comprehensive *agni*, but *ka* denotes the state within the said Hiraṇyagarbha specifically. It may as well be interpreted in a different way. *Ka* contains the first consonent + the first vowel. The combination makes it the letter abbreviating the entire language. Veda calls the creation a *kāvya* (poetry or verse) of the Creator. The *ka* indicates the material initiating such a *kāvya* and hence named as such.

The said sūkta is being reproduced below and discussed in detail. We begin with the first mantra already summarily discussed above:

hiraṇyagarbhaḥ samavartatāgre bhūtasya jātaḥ patireka āsīt
sa dādhāra pṛithivīṁ dyāmutemāṁ kasmai devāya haviṣā vidhema

Term meaning: (Hiraṇyagarbhaḥ) Already discussed at length above as the initial "pregnancy" of Prakṛti, so named by Veda, which grew into what the modern science calls a primordial fire-ball **(samavartata agre)** which came into existence in the beginning. **(ekaḥ patiḥ āsīt)** It was the sole master of **(bhūtasya jātaḥ)** whatever came to be born. **(sa dādhāra pṛthivīṁ)** It held within itself the earth-like planets **(dyāṁ utemāṁ)** and this entire stellar world. **(kasmai devāya)** For that illuminated one, the *ka*, **(haviṣā vidhema)** we offer our respectful oblations.

Translation: The mantra says that the very first transformation of Prakṛti that came to manifest was in the form of the primordial fire-ball. It was the master or possessor of all that was to manifest. Thc contents that subsequently produced the stellar world and the earth-like planets were solely held by it. To that which is called *ka* respectful oblations are offered.

Special Comments: The mantra specifically states that the contents of entire universe that came to be created later on were held within the Hiraṇyagarbha. Significance of such a nomenclature has already been stated above. The mantra implies that even we having come to be created out of the said contents, the *ka* deserves our homage which we provide by way of oblations. Offering oblations in the fire of a yajña is a token recreation at an extreme miniature level of the initial phenomenon.

ya ātmadā baladā yasya viśva upāsate praśiṣaṁ yasya devāḥ
yasya chāyāmṛtaṁ yasya mṛtyuḥ kasmai devāya haviṣā vidhema

(R.V. 10/121/2)

Term meaning: (ya ātmadā) One who is the provider of all motion and all knowledge; **(baladā)** who provides all strength and force; **(yasya viśva upāsate)** the entire universe is by whose side; **(praśiṣam yasya devāḥ)** whose administration or control is binding upon all; **(yasya chāyā amṛtam)** whose shadow or cover gives immortality; **(yasya mṛtyuḥ)** and annoyance death; **(kasmai devāya)** for that illuminated one, the *ka*, **(haviṣā vidhema)** we resolve, through the provision of these oblations, to serve its purpose.

Translation: The *ka* provides all knowledge, motion and force to universe which is at its side, and in control by its administration; life is sustained under its cover; and defiance means death. For it we resolve to serve.

Spl. Comments: The mantra underscores that all knowledge, dynamism and force in the universe are generated from that source '*ka*' whose working must not be defied, but rather studied and abided with.

yaḥ prāṇato nimiṣato mahitvaika idrājā jagato babhūva
ya īśe asya dvipadaścatuṣpadaḥ kasmai devāya haviṣā vidhema

(R.V. 10/121/3)

Term meaning: (yaḥ) Who; **(prāṇataḥ nimiṣataḥ jagataḥ)** of this biological as well as material universe; **(mahitvā)** on account of its greatness; **(ekaḥ it rājā babhūva)** is the sole ruler; **(yaḥ īśe asya dvipadaścatuṣpadaḥ)** who is the Lord of all the biped and quadruped world; **(kasmai.......vidhema)** For that '*ka*' we offer these oblations in token of our affirmation that we shall follow its rules.

Translation: The biological and the non-biological material world is born out of that *ka,* and is governed by its rules; this consciousness must dawn upon us. The two terms cover all creations.

Spl. Comments: The mantra says that the rules and laws set by that Supreme Master of the biological and the material world are to be followed. The repetition of the biological part in some detail led by the term *dvipadaḥ,* biped, expresses greater onus upon human beings for such an adherence (as this species is highly intelligent). That is why we resolve to dedicate ourselves wholeheartedly to the directions of *ka.*

yasyeme himavanto mahitvā yasya samudraṁ rasayā sahāhuḥ
yasyemāḥ pradiśo yasya bāhū kasmai devāya haviṣā vidhema
(R.V. 10/121/4)

Term meaning: (yasyeme himavantaḥ mahitvā) Having achieved greatness by the way displayed by its stellar world; **(yasya samudraṁ rasayā sahāhuḥ)** together with intervening space filled with energy, **(yasyemāḥ pradiśo yasya bāhū)** and these directions which are its arms; **(kasmai... vidhema)** we resolve, through these oblations for *ka* to endeavour to build up one functional system.

Translation: In the form of the entire stellar manifestation inclusive of the intervening space which is also filled with energy, as also different directions, *ka* has blossomed into an immensely large unified functional system. For its sake we resolve to work to that end in our world.

Spl. Comments: The mantra says that the entire greatness of universe seen in the expanse of the stellar world, the inner and the interspace and its expansion in different directions is the greatness of '*ka*' as one manifested system. We must put our hearts and minds to achieve such functional unity in our world.

yena dyaurugrā pṛthivī ca dṛḍhā yena svaḥ stabhitam yena nākaḥ
yo antarikṣe rajaso vimānaḥ kasmai devāya haviṣā vidhema

(R.V. 10/121/5)

Term meaning: (yena dyauḥ ugrā) That which made the stellar world fiery; **(pṛithvī ca dṛḍhā)** and the earth like planets solid; **(yena svaḥ stabhitam)** that which has held the burning bodies the way they are **(yena nākaḥ)** and the sun as well; **(yo antarikṣe rajaso vimānaḥ)** who makes bodies move birdlike in space; **(kasmai....vidhema)** for that '*ka*' through these oblations we resolve to unify our activities.

Translation: That which can generate intense heat in the stellar world, while solidifying earth-like planets, can make the sun hold its place in space the way it does, can direct bodies to move like spacecrafts, in short, can cause so many multifarious activities, functions in a manner having unified bearing within the universe, for such a '*ka*' we resolve to unify our activities in a harmonious and comprehensive manner.

Spl. Comments: Each one of the mantras in this Sūkta describes some characteristics in the universe which were possessed by *ka* while within the state of Hiraṇyagarbha and which after its release were to come to be manifested visibly in the universe in the form in which *ka* was to grow or develop itself. By mentioning such characteristics in tune with the process of creation the ṛṣi says the oblations better help us resolve to gain them in life. Obviously the advent of ṛṣis' was a very late phenomenon than the appearance of Hiraṇyagarbha, and they could realise the process of creation through a vision achieved in their presence.

yaṁ krandasī avasā tastabhāne abhyaikṣetām manasā rejamāne
yatrādhi sūra udito vibhāti kasmai devāya haviṣā vidhema

(R.V. 10/121/6)

Term meaning: (yam krandasī) Who was hailed aloud; **(tastabhāne)** by the substance of prospective universe; **(avasā)** with upbringing;

(abhi aikṣetām) imbued with the purposeful vision; **(manasā)** inherent within; **(rejamāne)** while in motion. **(yatra adhi)** Based upon which **(sūraḥ)** the sun and the stellar system; **(udito vibhāti)** having risen spread splendour. **(kasmai....vidhema)** We provide through the oblations to extend that glory all around.

Translation: The mantra communicates the emergence of the radiation following the big bang. It also mentions the purposeful vision for the creation of universe following which the stellar systems were to appear spreading splendour all around in space.

Spl. Comments: The Vedic term *dyāvāpṛthivī*, a compound of two words *dyū* and *pṛthivī*, denotes the entire universe. The same is understood in this mantra resulting from verbs etc., being used in dual number. Again the term '*avasā*' from the root '*ava*' declared by Pāṇini to cover nineteen realms of meanings, which with the use of prefixes and suffixes, may cover an immensely wide range of activity, has simply been mentioned to denote upbringing of the vast activity implied. It has already been clarified earlier that Puruṣa is instrumental in imbuing its wisdom related to the direction of creation at the *mahat* or *hiraṇyagarbha* stage itself, and through the radiation which subsequently roars out in a bang and proceeds straightaway with the business of creation, which has been indicated with the mention of sūra, the sun or the stellar systems in the later part.

āpo ha yadbṛhatīrviśvamāyan garbhaṁ dadhānā janayantīragnim
tato devānāṁ samavartatāsurekaḥ kasmai devāya haviṣā vidhema

(R.V. 10/121/7)

Term meaning: (āpo ha) It is definitely the quantum level activity; **(yaḥ)** which; **(garbham dadhānā)** caused conception to be held; **(janayantīḥ agnim)** producing agni, *i.e.*, radiation; **(bṛhatiḥ viśvam āyan)** capable to attain total cosmic manifestion. **(tataḥ)** From there; **(devānām āsuḥ)** for launching the particle; **(ekaḥ samavartata)** the solitary one existed. **(kasmai devāya haviṣā vidhema)** Oblations are provided to dedicate ourselves to seek sublimation.

Translation: The mantra says that the build up of *ka* must have commenced within Hiraṇyagarbha with Prakṛti getting activated at tanmātra (quantum) level, to create radiation and particles for cosmic manifestation. For launching the same the Hiraṇyagarbha existed. For the *ka* we resolve to universalise ourselves.

Special Comments: Recognising the expanse of *ka* i.e., radiation and particles in the universe and realising its source to be Hiraṇyagarbha, the ṛṣi states the initiation of its generation therein. The mention of *āpaḥ*, quantum level activity, is made very positive by the use of *ha* (definitely).

yaścidāpo mahinā paryapaśyaddakṣaṁ dadhānā janayantīryajñam
yo deveṣvadhi deva eka āsīt kasmai devāya haviṣā vidhema
(R.V. 10/121/8)

Term meaning: (yaścid mahinā paryapaśyad) Who with its pervasion saw from all around; **(āpaḥ dakṣaṁ dadhānā)** the particles giving birth to *'dakṣa'* **(janayantīḥ yajñam)** initiating the *'yajña'* of creation. **(yo deveṣu adhi devaḥ ekaḥ āsīt)** One who is the sole master over all those in motion, **(kasmai .... vidhema)** for that '*ka*' we provide oblations to achieve growth of our prosperity.

Translation: The mantra mentions the creation of *daksha* (deuteron) which is an initial step in the process of atomic creation. It also speaks of initiating the yajña, the process of creation of universe.

Spl. Comments: Veda speaks of *dakṣa* coming to be manifested within the Hiraṇyagarbha as part of *ka*. Modern Physics having come to accept the primordial fire ball as the source of emergence of cosmic process, has also come to accept the generation of deuteron within the contents of the said fire ball. *Dakṣa* is the Vedic name for deuteron as will be explained when the Vedic *prajāpatis* (sub-atomic particles) are discussed in a chapter to follow. It is also the top rung of the particle ladder representing the rest of them.

mā no hiṁsījjanitā yaḥ pṛthivyā yo vā divam satyadharmā jajāna
yaścāpaścandrā bṛhatirjajāna kasmai devāya haviṣā vidhema
(R.V. 10/121/9)

Term meaning: (yaḥ pṛthivyāḥ janitā) The one who produced pṛthivī, that is, life-bearing planets **(mā no hiṁsīt)** that the live bodies created by the same be not destroyed **(vā yaḥ satyadharmā divam ā jajāna)** or else the one who behaves to hold matter that created the entire stellar world **(yaḥ ca)** and who **(āpaḥ candrā bṛhatiḥ jajāna)** created highly potential and joy-causing tanmātras, *i.e.*, quantum level acitivity, **(kasmai ....vidhema)** for that *ka* we provide oblations in order to devise re-generation.

Translation: The *ka* created the fiery stellar world, where no life is possible, as well as the earth-like planets to bear biological creations. Such a delicacy of creative process was possible due to quantum level activity generated by it.

Spl. Comments: The mantra describes the contrast in creation. On the one side the fiery stellar world, on the other the life-bearing earth-like planets, both through the potential of joy-giving quantum level activity. The delicacy of the creation is being emphasised.

prajāpate na tvadetānyanyo viśvā jātāni pari tā babhūva
yatkāmāste juhumastanno astu vayaṁ syāma patayo rayīṇām

(R.V. 10/121 ends)

Term meaning: (prajāpate) O Lord of all the beings ! **(tvat anyat)** anybody other than you; **(tā etāni)** those and these, *i.e.*, distant and nearby located **(viśvā)** total **(jātāni)** born objects and beings, material and biological ones; **(na pari babhūva)** none happened who could transcend and control all, i.e., only you are capable of keeping this universe in order. **(yatkāmāḥ)** Desirous of whatsoever **(te juhumaḥ)** we offer oblations for your veneration; **(tat naḥ astu)** may that desire of ours bc fulfilled, by which **(vayam)** we **(syāma patayo rayīṇām)** become masters of riches for benevolence.

Translation: The ṛṣi says that in the entire creation, O Lord of all beings, there is none who can transcend and control the entire universe. We venerate you with oblations. Whatsoever we are desirous of that may be granted to us, so that riches belong to us for performing benevolent deeds.

Spl. Comments: Either in the earlier mantras wherein the expression used is *haviṣā vidhema* or in this last mantra where the verb employed is *juhumaḥ*, absolutely no personal desire is stated. All of these prayers, so to say, expressed are coined upon root *hu* to donate, to consume and to take or receive. According to Veda this is the universal manner of creation. Every object created is sustaining itself upon whatever it is receiving from others. That is why this root has come to be so deeply associated with yajña. Hence whatever is desired it is always along the process of creation which in fact amounts to earning for offering or donating back to the whole. Moreover, it is noteworthy that all such prayers are in first person plural number, and never in singular number. The Vedic desire for prosperity and riches is without exception, for all and by way of literally earning the said prosperity and riches by organising the working in the cyclic order of a yajña in which labour, wisdom or management, etc., may form contribution as oblations. The term *rayiṇām* a possessive plural of '*rayi*' further intensifies the spirit of the *yajña*. Derived from the root *rā* to give, bestow, impart in the dictionary (for the purpose to donate according to Pāṇini) *rayi* denotes wealth or riches meant for being donated or given away for benevolent objects or betterment of the rest.

It may further be pointed out that recitation of *kasmai devāya haviṣā vidhema* after each of the first nine mantras must not be taken as a mere repetition. The purpose appears to be to impress upon the human mind the role of *ka* in manifesting a wholly internally interdependent cosmic system based on each object donating its might for the whole, as well as to cause such a sentiment to be generated within the human beings towards the creation as oblations offered produce in a yajña.

— 3 —

AHAṄKĀRA

Sāṅkhya enumerating the various elements which caused the manifestation of the universe says: *"prakṛter mahān mahato ahaṅkāraḥ....."*, *i.e.*, Prakṛti caused *mahat* which in turn caused *ahaṅkāra*. *Mahān* and *mahataḥ* are simply variations of *mahat* in different cases. This *mahat* has been discussed in the previous chapter as the Hiraṇyagarbha of Veda or the state which developed into the primordial fireball of the modern science. From this secondary state emerged *ahaṅkāra*.

The term *ahaṅkāra* is an abstract noun of first person, singular number nominative pronoun *aham* in Sanskrit and *I* in English. Basically it denotes the sense of *I-ness*. It is, therefore, necessary to understand the delicacy and the propriety of this Sāṅkhya expression. The traditional or conventional meaning of *ahaṅkāra* is ego or conceit. But in fact it communicates the individuality of self, and it is this sense which covers the entire lot of manifestations collectively known as the universe, in which the same has been used. Sāṅkhya is only saying that all manifestations, without exception, emerged from *Mahat* possessing one fundamental characteristic of being unique in creation.

In the world of human beings each person uses the term I for his/her own self. This expression does not include anybody else, not even the dearest and nearest one. It is just the expression of exclusivity, the self of the person using it. This exclusive distinctiveness is the dominant characteristic of the creation, be it physical or biological. Each unit of creation enjoys its individuality among the whole mass of physical existence.

All human beings are basically similar; all animals of all species are but similar to each other among their respective categories. The same is the rule in the botanical world, nay, all physical manifestations whatever the size, shape or kind. But no two are the same anywhere. Every object, every creation, is a unique piece. It has similarities with so many, dissimilarities with a far greater number, but in its own right it stands as an original individual piece, without another same.

It is now an old story when it became known that the thumb impressions of two individuals are never similar. They do not match. We now know that not only the thumb, the impression of the formation of any part of the body would not match its equivalent in any other body. This, too, is old knowledge. Even the cells are not the same though they are very much similar. Each piece of the creative process, be it big or small, bears its originality. It has no duplicate.

Coming to the world of physics the scientific researches confirm this state of creation all over. The governing rule is the same. Things are similar but not the same. Even the atoms have an originality of their own. They don't have any thumbs to leave any mark. But the electron energy levels can be calculated. The spectra of the different atoms reveal their lines which indicate the energy spacings of respective atoms which are considered their characteristic individuality or we may say the 'fingerprint' of an atom by which it may be identified spectrographically.

Ahaṅkāra is the identity mark of creation. Sāṅkhya while presenting Vedic Physics relates this identity mark to denote the enormous multiplicity of material creation from the radiation and the particles that emerged with the bang of *mahat*. Nobody so far has summed up the creation so exhaustively in a single term since Kapila, the author of Sāṅkhya.

This uniqueness of creation extends to the entire process since its beginning till its conclusion. Two stars are never the same, neither is the period of their duration, nor ever prior to that nor ever thereafter. Neither any two planets nor any satellite planets like moon were, are or will ever be the same. The same is the state of being of all atoms, molecules or compounds. No two days are the same nor any two nights. In its long journey in time and space no location point is covered twice.

Each step covers original location and there is no halt. Until the point of time is not reached, each location belongs to the future, and the moment it is reached, it passes into eternity.

In short, the reach of *ahaṅkāra* covers the entire range of creation, its past, present and future. Hence Sāṅkhya completes its story in just three words: *Prakṛti, mahat* and *ahaṅkāra* bracketed together with Puruṣa heading this bracketed section. The entire multiplicity of creation has been covered. But Sāṅkhya makes a distinction between the process of creation and the matter created. As such it details the process of creation by enumerating the causatives of creation. These have two levels of working, the quantum and the relativity levels. These are covered in the following chapters.

— 4 —

ṚTA

Ṛtaṁ ca satyaṁ ca (*ṛta* and *satya*) were the first successive manifestations according to R.V. 10/190/1. In this sūkta, the story of creation has been described in a considerably abbreviated manner. It begins with the manifestation of *ṛta* and *satya*. Veda says that the materialisation resulted on account of the generation of rise in heat. To understand as to how this did occur, we analyse the two nouns *ṛta* and *satya* and see what they communicate. Let us take them in that order.

Ṛta is derived from the root *ṛ* meaning *gati*. The word *gati* mentioned here literally means motion. In Sanskrit grammar the biggest chunk of roots has been shown to mean gati. But the authorities have specified that wherever the meaning of any root has been so indicated the same will not be taken in the literal sense only, but shall be a pointer to cover three domains in its expression, they being *jñāna*, *gamana* and *prāpti* (knowledge, motion and reach or accomplishment) collectively or severally as the case may be. A root which is meant to cover merely any one or two meanings out of the said three, is not to be included in the *gati* group. The use of such a root or roots has been indicated by words other than *gati* even where the meaning is involved to denote literally motion.

Hence the root *ṛ* from which *ṛta* is derived, having been provided to mean *gati*, does cover the meanings related with any or all of the three domains as mentioned above. The derivation of *ṛta* is with the addition of a suffix *ta* to the said root. This suffix brings about the manifestation of the word *ṛta* in two different positions. In one case it is an abstract noun in neuter gender, while in the other it appears as a past participle. Our dictionary gives some of the meanings of *ṛta* as proper, right, apt, a settled order, law, rule, divine law, truth, in general righteousness, etc. It is noteworthy that none of these denote the sense of motion basic to the root.

As Veda states the emergence of *ṛta* to be initiation of the process of creation it is not far-fetched to infer that Prakṛti from the state of mere existence under its first attribute, has undergone a change by way of the second attribute *rajas*, motion, emerging and coming into play. In modern scientific language this would mean to say that energy at rest changed to kinetic form called K.E. Apparently the Vedic term for the more modern kinetic energy is also *ṛta*.

But the term K.E. does not carry meanings as quoted above in relation to *ṛta*. The reason is that the Vedic noun covers a far wider field of meaning and reveals as to how a superbly orderly system like universe came into being. It has already been pointed out earlier that Prakṛti of Veda and energy of the modern science are two synonyms of the same eternally present material cause of creation which is inert by itself, and resists change. But the Vedic term *ṛta* having been derived from the root *ṛ* also covers the meaning of knowledge as well besides motion. Hence *ṛta* does not merely mean motion, but reveals that its motion is charged with the knowledge of accomplishing some purpose. Now let us examine the meanings mentioned above. The *ṛta* is proper or right or apt to accomplish the state of manifestation for which it has been aroused to come under the influence of the second attribute *rajas*, motion, by terminating its earlier position of being at rest. On this basis consider the remaining meanings given above, and the *ṛta* will reveal the whole world of meanings connoting creation, maintenance, transformation in the most efficient and orderly manner in the greatest system known as universe. *Ṛta* is always 'right' not only in a literal sense in English language, but in the most comprehensive sense in any language, as the latter itself earns its origin from Sanskrit *ṛta*.

Ṛta is all science. *Ṛta* is divine. *Ṛta* is never wrong; it can never be wrong. It is always right. Anything that is in slightest divergence with *ṛta* is wrong. Vedic Sanskrit calls such a divergence *anṛta*, *i.e.*, not *ṛta*. The divine angle has been discussed and explained at length in the chapter on *Puruṣa*. Affected by such a divine touch Prakṛti in motion, *i.e.*, *ṛta*, invariably follows the creative process or plan and if the motion happens to take it stray it right back wheels into the correct process. Henceforth the discussion here will be confined to the scientific aspect only which runs along the mode of the kinetic energy of Modern Physics. It may also be added that the knowledge-bearing aspect of the *ṛta* has been discussed in the second chapter.

Ṛta through its constant motion lays down the pattern of the cosmic law which governs, controls and maintains all creation and the entire activity, actions and re-actions, related with the same. This cosmic law is called *dharma* by Veda. It must be very clearly and unmistakably understood that *dharma* is science and nothing but science, and most certainly not religion in the Vedic context. The noun *dharma* derived from the root *dhṛ* conveys the meaning of that which holds and maintains existence and continuance in both the universal and individual sense. Like any scientific position, there can never be two *dharma* on any given point or under any given circumstances. Like science, *dharma* can never be at cross purpose. *Dharma* is neither any worship nor any ritual. Science is the knowledge of the order and *dharma* is the order itself. That is why *dharma* is synonymous with the nature of the motion of things and their reactions, interactions. It is a different matter that in course of time conventionally it (*dharma*) was taken to mean religion, a meaning which stripped it of scientific wisdom. On the contrary, religion is the perception and interpretation given by an individual, howsoever saintly or great he might be. Because of his individual limitations he or she cannot equal Puruṣa, the Creator of *dharma*. The individual can at best perceive *dharma* and interpret it for others, but such an interpretation is open to examination.

Ṛta creates a field of its motion or activity for the creation of *satya*, i.e., matter. The noun *satya* means truth, but it also means reality. The very nomenclature is proof enough to reveal that Veda declares the material universe to be a real, vigorously active, superbly organized, wholesome system. Certainly it is no illusion, neither a dream nor an apparition of some kind, a form which does not exist in real terms.

Even in acient India, some of the giants gifted with razor-sharp intellect like Śaṅkarācārya, have gone astray at this point by taking it to a dead-end, instead of following the scientific track laid down by Veda. They have called this '*jagat*', universe in motion, as *mithyā*, unreal. It appears that the confusion arose on the interpretation of the term '*māyā*' which, due to constantly changing state of things and affairs in the material world, was conventionally got accepted to mean illusion. Creation of an illusion is *māyā* too, as the mind feels the apparition no less sensitively. But while trying to perceive a scientific line, it is the objective analysis or perception which pays, and not a subjective approach.

Māyā includes the knowledge to assume some form which may be subject to measurement. When some entity, which is formless and as such measureless, demonstrates the knowledge of assumption to some form subject to measurement, the gainful employment of such a knowledge is *māyā*. The dictionary uses the word 'wisdom' to state the said meaning. Knowledge and wisdom are synonymous terms.

According to Veda the manifestation of *māyā* is in the domain of the influence of the third attribute, namely, *tamas*, which has come to fore in respect of Prakṛti, the energy, part of which, beginning with the state of *sattva*, existence, having become mobile, under the state of *rajas*, motion, and finally moved into the top gear related to creation of matter by way of transformation into numerous forms, thereby bringing about the manifestation and maintenance under the process of an incessant change on account of its eternal motion. As stated earlier the knowledge that charged the conserved Prakṛti from its state of formless existence to the final state of transformation into material form is the play work named '*māyā*'.

When in early 20th century of the Christian Era, Albert Einstein declared that matter is energy transformed, the character of the study of Modern Physics changed and the whole of this universe became its field. Till then the study of Physics in the West was the study of energy and matter and their inter-action. But what Einstein brought into focus with his famous formula $E=MC^2$, thereby clarifying the finality of energy as the sole existence and matter merely its transformation, has all along been declared by Veda. This is no mean attempt to belittle the contribution of that great scientist to the world of Modern Physics. On the contrary, his expression of the scientific position by way of a mathematical equation proved his thesis to the hilt. What is here being pointed out is that the Einsteinian achievement in itself is proof enough of the authenticity of Vedic position.

Affected by the wisdom and direction of Puruṣa, energy recognised by the modern science as K.E., turns *ṛta* and proceeds to conceive, establish and develop as well as maintain after delivery the cosmic system of which we are though an infinitely minuscule yet a superb creation, at times even inquisitive enough to challenge or deny the very existence of all existences. But that is essentially the role to be played by the wisdom, to learn to know that chaff is merely a covering and not the grain, and it is the latter which is to be gained along with the knowledge regarding how to sow it, reap the crop and thrash it out.

The subject of *satya*, a Vedic term for matter of the modern science, will be discussed at length in quite a few of the chapters that follow beginning with the very next one. But before this chapter is closed it may be proper to explain here the mode of appearance of *ṛta*, *i.e.*, K.E. from *Mahat* or *Hiraṇyagarbha* after the big bang. That is why the entire sūkta R.V. 1/163 has been explained in a later chapter. But here we are concerned with the emergence of *ṛta* which has been described as **śyenasya pakṣā hariṇasya bāhū** in R.V. 1/163/1. The contents of this quote have been explained in greater details in chapter 8, wherein the entire sūkta has been discussed. It should suffice here to state that the said quote from the mantra states that the emergence of *ṛta* as radiation occurred in two forms, the electromagnetic waves and the matter waves, the latter being also known to the modern science as the de Broglie waves. The **śyenasya pakṣā** part indicates the former and the **hariṇasya bāhū** the latter of the above two. But let us now move on to *satya* first.

— 5 —

SATYA

As quoted earlier **'ṛtaṁ ca satyaṁ ca'** (R.V. 10/190/1) *ṛta* and *satya* were the successive manifestations of *Prakṛti* after the commencement of the process of creation. *Ṛta* has already been discussed earlier. Let us see what the other noun *satya* has to say for itself. This Sanskrit term is conventionally considered to have as its synonym the noun *truth* in English. But in physics it denotes reality. Its etymological analysis will reveal its correct interpretation. The noun *satya* is derived by adding suffix *ya* to *sat*. We may recollect that the term '*sat*', already discussed in relation to the first attribute of *Prakṛti* in chapter 1, denotes the state of being eternally present or existent. The rule of grammar under which suffix *ya* is added states that the noun so derived, that is *satya*, shall denote the product so named to be in virtuous harmony with whatever is *sat*. Veda has stated *ṛta* and *satya* to be the succeeding manifestations and all manifestations are from *Prakṛti*, which is *sat* by virtue of its first attribute, or the *ṛta* which is as well *Prakṛti* itself so named under *rajas*, motion, the second attribute. The term *satya*, therefore, denotes its state of being under the third attribute, namely, *tamas*, i.e., transformation as mass, matter or any other physical form howsoever fluid. We shall use the lone term *matter*. In other words Veda states that matter is *Prakṛti* itself so transformed in its third state of being and that the state of matter is in virtuous harmony with *Prakṛti* being but a transformation of the same. Any annihilation of matter will cause it to revert to its earlier state of being *ṛta* or *Prakṛti*.

Modern Physics tells us that since the days of Aristotle, or maybe even prior to him, physics has been considered to be the study of energy and matter and the interaction between the two. It took the arrival of the 20th century and the appearance of an Einstein to establish that the mass or matter was nothing but transformed energy and with his famous equation $E=MC^2$ to change the entire thrust of research and study in the field of physics.

Vedas have been stating this scientific position from the day one. The choice of the terms *ṛta* and *satya* so eloquently speaks that the latter is but a transformation of the former. It is also clear from the first chapter on *Prakṛti* that it remains the same in any of its three states of being, i.e., the state of conservation, and that it is the material cause of creation. Veda further clarifies, as we shall see later, that the involvement of *Puruṣa* does not go beyond providing dynamism and direction to motion through creative process. Thus the fundamental material cause of all creation rests in Prakṛti while the supreme wisdom manifest in the birth, growth and maintenance of the cosmos emerges from the efficient cause. The conclusion is obviously that the entire material or physical part of existence is but *Prakṛti* in any of its three states.

One may point out that Einstein laid down a formula. Do we find the same in Veda? Not specifically. But the particular formula and all the formulae supporting the universal structure and functioning are part of the wisdom and not matter which is inert. Veda in the form of Yoga does reveal a discipline which helps an individual human being to arouse his or her intuition or superior powers of perception for solving the problem or through the advanced state of Samādhi have a vision of the Supreme ocean of wisdom encompassing the universe and beyond. Vedas themselves state that they have been revealed for human beings. Even the history of the growth and development of the modern science bears testimony to the fact of intuitive findings coming to the rescue of so many haggard scientists. The intuition provides a glimpse of the solution, while the yogic samādhi stabilizes the vision.

It may, even at the risk of stepping beyond the subject under discussion, be pointed out with reference to Sanskrit *satya* being synonymous to English term truth that truth is without exception supportive of any system that stabilises its existence. Hence truth by itself is ever in virtuous harmony with existence, i.e., *sat*. In a different context even Veda holds truth to be in the mould of the Supreme Being, as explained in Chapter 18.

By naming matter as *satya* Veda emphasises that the total cosmic system is fundamentally one entity. The created universe is one organisation in toto; only a solitary existence is manifesting itself everywhere. Right from the atom and the particles smaller than that to the bodies that are great and the ones still greater, in each body the *ṛta* is dancing. In *Nirukta*, the authoritative treatise of Yāska for the interpretation of Veda, *ṛta* has also been stated to denote electricity. The negative charge of particle electron and the positive charge of proton are its two legs or wings. Modern science considers the said charge of the two particles to stand neutralised in a neutron. The two charges do not become non-existent. They just balance each other. The atom itself in the entire growth process of its creation, similarly keeps the two charges balanced. The entire process of creation is ruled by the total dominance of energy. But the state of equilibrium or balance provides stability. The charge exists and yet is not operative, being altogether balanced. Such is the manifestation of *ṛta* in the constitution of matter.

According to the Vedic wisdom the journey from *ṛita* to *satya* gets completed in five stages. Each stage is named a *mahābhūta* which emerges from its quantum level. All of these five *mahābhūhtas* are causatives effecting *ṛta* to be transformed into *satya*. At the end of the process *ṛta* itself gets stabilised in the form of *satya* which is known to the modern science as matter.

Like the group of three particles, namely, electron, proton and neutron, forming the basic structure of atomic constitution, *satya* as matter puts in its appearance based on a group of three formations, i.e., an atom, an ion and a molecule. Molecule is the final form. These three get together through bonds and produce bigger molecules which further develop tie-ups. Their combinations in the form of compounds, etc., and finally forming the bulk of one or the other celestial bodies is a natural consequence of the dynamics of creation. During the life-span of the universe such activity goes on non-stop at various levels.

All over the universe the creation of *satya* or matter continues to be effected through a uniform five-stage process. The process itself is *satya*, rather each part of that process is *satya* and the product called matter by the modern science, delivered through it is *satya*. Simultaneously, whatever else that gets produced like anti-matter is *asat* in our universe. But in short the entire system is called satya. The reason is that the totality of the entire working system is associated either with the motion of *ṛta* or is a result of its transformation. Whatever exists at the material or physical level is a transformation of *ṛta* one way or the other. The entire *satya* falls under the wings of *ṛta*. It is the *ṛta* which is existence in motion.

The creation of universe has a system of its own. Under the discipline of the said system every single body or group of bodies, each one of the stars or constellations, the group of stars or the galaxy, is in uninterrupted motion along its orbital track. The possibility of collision occurring among them is little. The number of accidental collisions in any modern metropolitan city's highly organised traffic system upon the surface of this earth ranges on fairly high side than such occurrences at cosmic level. On account of such a system the universal activity has been going on smoothly, and is expected to keep on running the same way. This system is named *dharma* by Veda. The said system or *dharma*, being virtuously homogeneous in *ṛta* is also *satya*.

This cosmic system is in evidence right from the tiniest particle to the greater than the greatest of body systems and the expanse. In fact the entire system of creation is the result of the motion of *ṛta* or K.E.. The entire activity is occurring in harmony and in accordance with the same. It covers the entire range of the biological as well as the physical creations of whatever stature and magnitude. The said motion is *sat* and whatever activity is occurring, being in accordance with same is *satya*. The matter is a product of *ṛta*, its very existence being based upon the latter. Such a basis provides matter its indestructibility because it is the *ṛta* which, by way of impact, has bound itself under measure or limit. That is why in spite of destruction caused to a body, its matter conserves itself through some other body form. In Sanskrit language, destruction is called *nāśa* or *vināśa*. In the latter term *vi* is merely a prefix to emphasise the noun *nāśa* which is derived from the root *naś* to disappear. Veda says all destruction is simply disappearance, not annihilation. It is the form that gets destroyed or annihilated not the substance which merely disappears. Even when there is total annihilation as in the case of a matter and an anti-matter particle collision, the annihilation of the two is in fact disappearance of their respective physical forms; the energy involved in their manifestation remains conserved in its release. But Veda makes another scientifically potent statement regarding *satya*:

hiraṇmayena pātreṇa satyasyāpihitaṁ mukham

(Y.V. 40/17)

It means that the face of *satya* is veiled under the cover of light. This is the forepart of the last mantra of Yajurveda. To clarify what is being communicated by the mantra, it would be useful to have an illustration. Let us imagine a railway track with two persons standing near it on the same side at considerable distance apart watching a train moving between them from the side of the one to the other. To the person watching it going away, the train would continue getting smaller and smaller, while the one watching it approach would see it getting bigger and bigger. The train is the same as it is. Why does it appear to be so? The difference is due to the picture that is being formed upon the respective retinas of the two observers, in one case that of the train going away, while in the other that of coming nearer. The fact is that each one of them is in fact seeing the picture as received upon his ratina, although he honestly feels that he is viewing the train itself. We see what our ratina shows to us, i.e., our mind, but it appears that we are seeing the real thing or event. None of the two, or for that matter, no human being sees the *satya* (the reality of manifestation). We all see the picture that gets reflected upon our retinas. The phenomenon is due to relativity.

The phenomenon is all the more noteworthy in respect of colours. The natural light from the sun consists of all the colours of the spectrum. It drenches the world outside with all of them. But we see a rose red or yellow or black as the case may be, because the flower has absorbed all other colours and is reflecting the particular colour or colours only to our ratinal reception. Hence nobody sees the real colour of the rose, or its seven colour drenched image. Only what is being received is being communicated to the mind. The same is true of a colour camera where the film is receving the reflected light. This optical illusion, so well known to the modern science now, has been stated by Veda. The mantra partly quoted goes on further to say that recognising *satya* out of the appearances is the real knowledge or science.

Another Vedic declaration regarding *satya* is worthy of consideration as well:

satyameva jayate nānṛtam

It says: *Satya* is finally victorious; what is in disharmony with *ṛta* is never. Victory is gain, in whatever field it is achieved; victory is life. If the same is being sought along a path that is in harmony with *ṛta* the endeavour is bound to succeed finally whatever be the odds. *Ṛta* as explained earlier, denotes three aspects, namely, knowledge, motion and accomplishment. *Satya* being the manifestation of *ṛta* is endorsed by the three. Knowledge denotes total knowledge related to the purpose sought to be accomplished including the plan of action; motion is resolute action according to the plan; and there can hardly be any doubt about the victory which is the accomplishment of the purpose. *Ṛta* is Pr̥akṛti put in motion by Puruṣa under a plan conceived with the application of totality of knowledge that It is, to achieve creation of universe, and the victory or success is there for all to see. Moreover the verb in this quote above is *ātmanepadī* that is, passive voice. But in Sanskrit language such a verb conveys its total benefit for the sake of the self or the subject, and none for anybody else. As such the verb *jayate*, i.e., being victorious stands eternally with *satya* ultimately and not in any case otherwise. Hence it communicates that betterment remains always with the cause of *satya* or the creative process.

Reverting to the field of physics, it has already been pointed out above that *Prakṛti* under its third attribute *tamas* has been named *satya* by Veda as under the second attribute it has been called *ṛta* while under the first *sattva* it is known under its original name. The *satya* has materialised in numerous forms which may be broadly categorised in three states: the particle state, the atomic state and the galactic state which includes the planetary state as well. These will be presented in quite a few of the following chapters.

— 6 —

BRAHMĀ AND PRAJĀPATIS

Who is Brahmā? For a proper presentation of Vedic Physics let us modify the question properly. What is Brahmā? A straightforward answer to that question is: Brahmā of Vedic Physics is the atom of Modern Science. It has already been pointed out that for a correct understanding, the Vedic terms are to be interpreted on the basis of their derivation and with reference to the context or the subject. Here we are discussing the presentation of the Science of Physics, and let us see what the term *Brahmā* has to say.

The noun *Brahmā* is derived from the root *bṛhi* which means to grow, increase, expand, etc. Hence the basic characteristic of *Brahmā* is to grow greater and greater. So is that of an atom which puts in its appearance as a hydrogen unit and grows to greater and greater forms as shown in the Periodic Table. Brahmā is said to be the creator of universe, and so is an atom. The universe has been created of atoms. Brahmā is again said to have carried out such a creation out of its own body. This is fully true with regard to atom. The body structure of an atom consists of three sub-atomic particles, namely, electron, proton and neutron. They are in use everywhere. Only the first atom, that of hydrogen, has two of them, an electron and a proton. This part would be more elaborated when Vedic *prajāpatis* are explained below in this chapter. Finally the last but not the least characteristic of *Brahmā* is its being *caturmukha* (conventionally taken to mean having four faces). The term *mukha* in the compound above, is derived from the root *khan* meaning to dig, to tear. The addition of *mu* in the beginning and the removal of *n* from the root *khan* is done with the application of the suffix *da*, which also disappears. We have the noun *mukha* which means that which digs or tears. An ordinary student of physics learns that the increasing number of electrons in the orbit, as the atom grows with the inclusion of more protons (and neutrons) in the nucleus, seek accommodation in sub-orbitals which are not more than four and are known by the first letter of their respective names as s,p,d and f. The eletrons orbiting these four sub-orbitals do appear to be tearing the atomic space. *Catur* the first part of the compound above is just four. Hence atoms are *caturmukha*. This final congruence establishes *Brahmā* as the atom of Vedic Physics. A few other minor details are being left out but they all strengthen the same conclusion.

We may now consider the names of *prajāpatis* connected with *Brahmā*. They are listed below. The first seven are also known as *saptarṣis* (seven Ṛṣis), the 8th is called *devarṣi* (luminary Ṛṣi) and the last two happen to be the first predecessors of their respective successors. All of these are known to the science of physics and modern synonyms are given below against the respective Vedic names.

Veda		**Modern Physics**
1. Marīci	=	Photon
2. Pulaha	=	Neutrino
3. Vasiṣṭha	=	Electron
4. Atri	=	Muon
5. Pulasti	=	Pi Meson or Pion
6. Kratu	=	Proton
7. Aṅgiras	=	Neutron
8. Nārada	=	Pion (Neutral)
9. Bhṛgu	=	Plasma
10. Dakṣa	=	Deuteron

Lct us examine the meanings contained within these Sanskrit nouns through their etymological interpretation. We take them one by one:

1. Marīci: Derived from the root *mṛ* meaning to die, *marīci* is said to be the seniormost offspring of Brahmā the atom. Our dictionary states it to mean a particle of light which a photon is. Still we pursue this comparison a little further. A photon is a quantum of electro-magentic energy emitted by an electron. Being a quantum particle means that it is sort of a packet of energy/light and is unable to carry any more or less. It dumps its entire content wherever it lands and that is the end of its own existence. Also being the first among the elementary particles, and having no mass, it *ipso facto* heads the list. Veda has choosen the root *mṛ* to denote its characteristic behaviour.

2. Pulaha: The noun pula derived from the root *pul* meaning greatness or largeness, is joined to root *han* meaning to kill. Its ending n having been removed with the addition of suffix *ḍa*, which too disappears, this noun denotes one devoid of any greatness/largeness. It is a being which is too small for the purpose of any measurement of its mass. The Vedic term appears to be more self-explanatory than the name neutrino currently in vogue in Physics.

3. Vasiṣṭha: Derived from the root *vas* to reside this name or noun is formed with the addition of the superlative suffix *iṣṭha*. It denotes that the bearer of this name enjoys ą residence from which dislodging it is not easy. On the contrary it is extremely difficult, and may require tremendous force. We know that an electron enjoys an orbital residence around a nucleus. We also know that the first atom, i.e., the hydrogen atom enjoys stable existence. The state of stability means existence of one of the strongest bonds between the rotating electron and the proton, the lonely nucleon, so aptly denoted by the use of the superlative suffix in this Vedic term. Hence Vasiṣṭha is the Vedic name of an electron.

4. Atri: *Tri* in Sanskrit is three in English language. The *a* in the beginning denotes denial. A+tri= not + three. Hence the name *Atri* means that which is not three. In Sanskrit we would say '*na traya ityatriḥ*' or not three, that is, *atri*. This suggests some sort of a situation that appears to exist wherein the ghost of three prevails but is being denied in order to avoid any confusion. Modern Physics tells us that a muon decays into an electron, a muon neutrino and an electron anti-neutrino. But physics treats it as a single particle. Well, Veda is already decrying the existence of the count of said three through its nomenclature.

5. Pulasti: Once again we have the root *pul* meaning greatness/largeness in action here. The addition of the suffix ti to the noun *pulas*, derived from the said root, denotes that the object denoted by it possesses greatness/largeness in some consequence which imparts it a characteristic distinctiveness. We learn from Modern Physics that the pi meson or pion is the largest particle in size among the negatively charged particles. In other words, out of the neutrino, the electron, the muon and the pion, the last is the biggest among them. This identification so well established through its name declares Pulasti to be the pion of the physics of today.

6. Kratu: The noun *kratu* means a *yajña* or a doer/performer of *yajña*. We presumed that at least the readers of English language having an Indian background would know what a *yajña* is, but even such a presumption is doubtful. We have so far, in spite of our efforts, not been able to find any synonym or a near-about of *yajña* in English. As *yajña* is a word that carries in both physical as well as spiritual sense the very fundamental idea behind the creation of universe, it is the kingpin around which rotates the entire theme of Veda. To translate it as sacrifice or sacrificial fire is putting it very very mildly. That is why instead of translating the same we have preferred to use the original Sanskrit term *yajña* itself. We may return to an elaboration of *yajña* in a later chapter on *yajña* itself. Here we return to Vedic particle physics.

As *kratu* is a *yajña* or its doer/performer, one of the activity related to the same is *saṅgatikaraṇa*, i.e., unification of two or more actions in a single composite activity. Such a unification is *yajña*. We observe that in an atom the activities of a proton as well as an electron are so unified that each has a bearing on the other. Placed in the central position, the nucleus, the proton, happens to be the performer of such a unified activity. That is why the proton is the Vedic particle *kratu*.

7. Aṅgiras: Derived from the root *aṅg* meaning gati covering the field of knowledge, motion and accomplishment it is born, according to Nirukta, out of intensely burning particles. **aṅgāreṣvaṣaṅgirāḥ (*sambabhūva*)**, *Nirukta* 17/3. The bracketed part is ours to clarify the sense of the referred quote from the treatise on Veda. It means that *aṅgiras* were born as stated above.

Modern Physics tells us that the neutrons were born within the primordial fireball, the Vedic Hiraṇyagrabha discussed in chapter 2. The temperature has been estimated to have risen to beyond 1000 billion K. That is what quotation from *Nirukta* is suggesting. In short *aṅgiras* is the Vedic name for the same.

8. Nārada: Derived from the root *rad* meaning to divide, this Vedic term is a double negative compound, a unique etymological operation in Vedic language to emphasize unwaveringly the meaning contained in the verb root under operation. Out of a number of meanings conveyed by the root *rad* we pick up to divide from our dictionary to focus attention on this particular derivation. As stated above this is a double negative compound, the first being *a+rada* = *arada* meaning *no+division* and the second *na+arada* meaning *no+no division* or a most certain division. And that is what the term Nārada communicates. It names the particle which displays such a trait of division and yet is considered a single undivided independent particle.

Such a particle known to the physicists is *pion* (neutral) which enjoys a very short duration of life after which it ends up by dividing itself into two photons. It may further be pointed out that all elementary particles mentioned above from 1 to 7 have their reflected images as respective anti-matter particles. In such a reflected image the poles of the matter particles get reversed, negative particles like electron becoming anti-matter positive positron and the positive proton an anti-matter negative proton. The neutron has an anti-matter neutron. The Vedic *Nārada* or pion neutral goes through no such reversal, the anti-matter pion being there as a reversed image of positively charged one. The characteristic of a *Nārada* is that it ends up in two photons. Hence it is a solitary particle of its own kind.

For a short while we leave our reference of the subject of anti-matter particles to which we shall soon return before the conclusion of this chapter. It has already been mentioned that the *Prajāpatis* numbering 1 to 7, are also known as *saptarṣis* (seven ṛṣis), and the 8th *Nārada* as a *devarṣi*. Let us look at this aspect first, so as to complete our consideration in respect of the first eight. The remaining two fall in two separate physical identities.

The term *ṛṣi* is derived from the root *ṛṣi* denoting *gati*, i.e., knowledge, motion and accomplishment. A *ṛṣi* may well be a person of vast knowledge. That is why *Nirukta* has defined *ṛṣi* as a *mantradraṣṭā* (seer of Vedic mantras or any hidden knowledge). The Vedic text contains names of numerous *ṛṣis* who were the seers of a mantra or the group of mantras that follow their respective names. But any physical object, which is in motion and possesses the capacity to reach some goal by way of transformation, merger or even otherwise through its motion/action, displaying some systematic process, may as well be named a ṛṣi in Vedic parlance. For that matter the Sun is a *ṛṣi* and so is the Earth. The stars are all ṛṣis, as are the elementary particles discussed above. The seven stars which appear to be going round the Pole star in the northern sky have also been named with these very same Vedic nouns. The whole group of seven is also called *saptarṣi*. But this is a separate identification, which must have originated when Vedic scholars started using Vedic terms for the purpose of cognizance in human behaviour. Adoption of such a practice is on record in Vedic literature and *Manusmṛti*. As the revelation of Vedic text preceded its terms being put to use elsewhere, these seven names denote the elementary particles referred to from 1 to 7 above. The application of these very names with specific stars and/or human beings is a subsequent happening.

The 8th Nārada, solitary neutral figure, probably came to be addressed as *devarṣi* because the first word *deva* means a luminary and the particle, practically within no time, transforms itself into a pair of gamma rays, elctro-magnetic radiation. It also has no pair by way of an anti-matter mirror image.

Let us now take up the remaining two *Prajāpatis*:

9. Bhṛgu: Derived from the root *bhṛsj* to roast the noun denotes that which is well roasted. This characteristic is a clear indicator towards plasma the only sub-atomic state of matter which exists inside our sun and other stars in a constant state of being roasted. Albert Einstein has called it the fourth state of matter besides the three, solid, liquid and gas, available on our planet. This name must not be confused with the famous human dynasty associated with it. As a *prajāpati* this falls totally in a different class and denotes the causative state of matter.

10. Dakṣa: Derived from the root *dakṣ* meaning to grow, the *Dakṣa* as a *prajāpati* is that initial state of atomic growth which is known to physicists as a deuteron, the nucleus of first isotope of hydrogen, also called H_2. A deuteron is a heavier nucleus on account of the addition of a neutron which gives company to the lone proton already there. In our discussion regarding the 7th *prajāpati*, *Aṅgiras*, which is the Vedic name of the modern neutron, it has been clarified as to how they were formed. These neutrons are neutral in charge. They are responsible for bringing about merely a physical change in the state of growth of the relevant atom and restraining any chemical change in its nature. That is why their addition each time to the nucleus produces an isotope. The word *dakṣa* also means adroit or expert which is justified by the results of the process initiated by it at its level, that is, initiating the growth of atom.

It would be appropriate to clarify here that the noun Dakṣa denotes merely the state of fusion of one neutron with a proton, which is called a deuteron by modern science. It is the nucleus of the isotope H_2, also called heavy hydrogen. The isotope H_2 is also called *dvyaṇuka* in Veda, literally meaning a state accomplished by the fusion of the mass of two hydrogen atoms. Similarly the Vedic noun for a tritium or a H_3 isotope is *tryaṇuka*, i.e., a product of the fusion of three hydrogen atom mass. Veda has specified separate names for the different states of a deuteron and a deuterium and another for a tritium. Modern science, too, stands in the Vedic shoes. It has not named any other subsequent isotope as such, and it takes cognizance of a deuteron and a deuterium separately as two states of creation.

A few words to explain the term *prajāpati* here would also be in order to have a clearer understanding of Vedic position. The word *prajā* means procreation, offspring, aftergrowth, etc. The term *prajāpati* as a whole means a procreator. All of the 10 mentioned by Veda as *prajāpatis* in the context are procreators of matter in atomic or other subsequent forms we find in the universe. The 9th one *Bhṛgu* and the 10th one *Dakṣa* are not mere single particles by themselves, but are both the creators of respective processes of creation. The reason for first 8 coming to be included in the list of elementary particles, probably is the smallness of their size. That is what the word particle denotes. But Veda presents them by way of a class or category based upon their role of procreation.

There are many more particles that come to be created and end up in decay. Modern Physics mentions knowing about 200 particles. Veda gives a count of three times sixty that decay and our dictionary bears this reference against *marut* (R.V. 8/96/8). The whole group has been named *marut* or *marudgaṇa* (*gaṇa* meaning a group). Such a name betrays their nature of decay. Derived from the root *mṛ* meaning to die or decay, the noun *marut* comes into being with the addition of the suffix *ut*. It means that which dies or decays. Veda (R.V. 8/96/8) says ***triḥ ṣaṣṭistvā***. It means maruts, you are three times sixty.

We conclude this chapter with the presentation of the Vedic picture, regarding anti-matter particles. It has already been stated earlier that according to Modern Physics, anti-matter particles are reflected images of matter particles. The reflection provides them identical image but with the position of poles and consequently that of the charge reversed. A negatively charged matter particle generates its reflection as positively charged one, a positive one as negatively charged unit and the neutral neutron as neutrally charged reflection. Physics has named the anti-matter electron as *positron*, *posi* in forepart of this noun indicating its positive charge. The rest of the anti-matter particles are addressed by their matter names with the addition of anti-matter prior to them as an adjective.

The most important characteristic of an anti-matter particle is that any meeting between a matter and an anti-matter particle results in an instantaneous total annihilation of the two and their mass achieving hundred per cent transformation into energy. Whenever and wherever a pair meets its whole existence changes to energy. Only photon is an exception to this annihilation. Even with the poles reversed the photons survive. But then, a photon is itself energy. It has no mass, whatsoever, which may get annihilated.

The Vedic noun for anti-matter is *Kardama*. It is also a *prajāpati* but in its own right. The famous Sanskrit dictionary, *Śabda-kalpadrumaḥ*, states that *Kardama* is *brahmāṇas chāyāyāṁ jātaḥ*, i.e., born from the reflection of Brahmā. Brahmā is atom. The same origin of *Kardama* has also been given in our dictionary. The origin from *chāyā* (reflection) clinches the issue. It denotes anti-matter creations.

The following is the manner in which the Vedics pair their matter as well as their respective anti-matter particles, all of which bear a name:

	Matter	Anti-matter
1.	Marīci (Photon)	Sambhūti
2.	Pulaha (Neutrino)	Gati
3.	Vasiṣṭha (Electron)	Arundhatī (Positron)
4.	Atri (Muon)	Anasūyā
5.	Pulasti (Pion)	Havirbhū
6.	Kratu (Proton)	Kriyā
7.	Aṅgiras (Neutron)	Śraddhā

Their etymological meanings may also be considered:

Sambhūti: It is an abstract noun from the root *bhū* with the addition of suffix *ti*, the term *bhūti* meaning existence. The placement of prefix *sam* denotes their coming into existence together. Thus *sambhūti* means coming into existence together with its matter particle *Marīci*. This is a must in case of any reflection. It can never occur all by itself.

Gati: This noun virtually means velocity, which is a characteristic feature of pulaha. Its velocity is the same as that of light.

Arundhatī: This is a compound word. The first part *aru* is derived from the root *ṛ* meaning *gati*, which includes the sense of accomplishment. It also denotes *vraṇa*, a wound. But the root *vraṇ*, out of which noun *vraṇa* has been derived, means to destroy the body. Hence the meaning *vraṇa* would imply destruction of the body. The second part, *dhatī* being derived from the root *dhā* to inflict giving the meaning of an inflicter; the compound word comes to mean an inflicter of the destruction of body.

Anasūyā: A negative compound from the noun *asūyā* meaning grumbling (while performing). Negatived the name means not grumbling (in its performance) or ever ready. The performance has been abundantly made clear in the preceding interpretation, i.e., instant annihilation.

Havirbhū: This is also a compound of two words *havir* and *bhū*. It simply means born from *havir*. What is *havir* or *haviḥ* in any atomic composition? In an atom the nucleus is governed by the strong nuclear force. There are only two particles known to stay in the nucleus, proton and neutron. How do they stay? They stay together through strong interaction. The process of such a strong nuclear interaction is conducted through the exchange of the particle pi-meson back and forth like a ball being exchanged between two players. The word *havir* is derived from the root *hu* meaning to give and take. It may be used for giving and taking, i.e., to exchange. The noun *havir* derived out of the said root means that which is given and taken or exchanged in a *yajña*. It has already been pointed out earlier that the Vedic view about an atomic activity is that of the performance of a *yajña*. All this discussion brings us to the conclusion that the particle pion is the *havir* of the *yajña* being performed within an atom. Out of such a *havir* the *havirbhū* is born (through reflection), *bhū* meaning born.

Kriyā: Both the nouns *kratu* and *kriyā* are derivations of the same root *kṛ* meaning to do, to perform. Naturally one reflects the other.

Śraddhā: Derived from the root *śrā* meaning to burn, roast, the word '*srat*' as it appears in this compound means that which has withstood or survived the intense burning or roasting. The root '*dhā*' out of which the second part is derived, apart from what has been stated above in respect of *Arundhati*, also denotes to be held. In the light of this analysis the word *śrad* indicates a neutron and *dhā* its image, which reflected and held by it causing the manifestation of the anti-matter particle named here.

This adds up the presentation of Vedic particles to 197 which is about 200. There may still be a few more. Our dictionary states one marya as a particle. This name would also be derived from root *mṛ* to die or decay. It probably indicates the Lamda group of particles which decay.

— 7 —

A UNIVERSE IS BORN

Veda states an eternal cycle of creation and dissolution (*sṛṣṭi* and *pralaya*). One follows the other in an infinite sequence with no beginning and no end. The creation is a multiple manifestation of Prakṛti in a homogeneous, smooth orderly organisation called universe which after its duration, whatever it may be, merges itself into Prakṛti, which remains conserved in its formless inert existence. This state of existence is by far no end. The cyclic system goes on in eternal motion with the initiation of a new creation.

In short the present creation is not an initial creation. It is simply one in an infinite chain of creations and dissolutions. There was a dissolution before the present creation, which was preceded by an earlier creation and so on till such a continuity is lost in infinity. And the present creation is to be followed by the great dissolution in a constant cycle of their alternate occurrences along a spiraling continuum in infinity of time.

As any continuity presupposes its existence based in time, Veda indicates such a situation in the opening quarter of the first mantra of R.V. 10/129 which says **'nāsadāsīnnosadāsīttadānīm'** by the use of term *tadānīm* meaning at that time as pointed out by Dr. Mahavīra in chapter 19. The time gets unified with the state of Prakṛti in existence: so says Vaiśeṣika. But during the creation it emerges as the eternal measure both by way of count and duration to reveal the infinity of existence, wisdom and creativity as the properties of the Absolute.

Modern science holds the process of creation to have started with a big bang caused by an intensely blazing primordial fireball blast at a temperature beyond 1000 billions K. The blast caused the radiation and other contents within the fireball fly all around in space. Veda as reported in the second chapter narrates this in a slightly different manner that all the earths and all the stars were formerly held by the '*Hiraṇyagarbha*' within itself. In other words, the building material of the universe was held within it.

According to Veda Prakṛti or energy is the material cause of universe. But it is inert by nature. It exists, and exists eternally. It cannot be created and it never gets destroyed. Under any circumstances it remains conserved. On account of its inert nature it cannot even stir on its own. not to say of creating a universe. But an external cause can make it take to motion, its second attribute. Its third attribute is that through motion it can become compact or solid. The creation of universe is based upon these two of its attributes apart from its existence which is eternal and fundamental in any case.

Veda, therefore, declares that there is an efficient cause which is instrumental in making Prakṛti operate and deliver a universe. Veda names it *Puruṣa*. This is totally non-material. Like Prakṛti the existence of *Puruṣa* is also causeless. It is all wisdom and dynamism. It exists everywhere with no exception and over-extends the existence of *Prakṛti*. Sāṅkhya enumerates it last in its list of 25 elements. *Puruṣa* enjoys the finest state of being.

Well, how does It do it ? Sāṅkhya says: **uparāgāt kartṛtvam citsānnidhyāt**-1/164. It means that in juxtaposition with it Prakṛti was influenced to take to the state of being the performer. The term '*cit*' denotes wisdom and dynamism. It has been used for *Puruṣa*. In a later chapter this would be discussed at length. In the state of juxtaposition, when Prakṛti was at rest, Puruṣa caused a *kṣobha,* an agitation or a disturbance in the former.

What was this *kṣobha* or the agitation so caused ? The indication is available in the first half of the mantra R.V. 10/190/1:

ṛtaṁ ca satyaṁ cābhīdhāt tapasodhyajāyata

It says that *ṛta* and *satya* were born successively from **iddhāt tapasaḥ**, i.e., heat increased intensely. So the *kṣobha* referred to above was caused efficiently in the state of its tapas, that is, heat cum temperature. In the state of being at rest *Prakṛti* would have been at sub absolute zero degree temperature. How much below, there is no way of knowing. But the *kṣobha* effected rise in temperature. As soon as it passed the absolute zero point it effected generation of motion in the energy at quantum level. The process of creation had commenced.

If we may use a biological illustration to explain this event, it may be said in somewhat metaphorical terms that the *kṣobha* generated by *Puruṣa* was sort of implanting a spermatozoon of a universe, and *Prakṛti* having conceived it. The subsequent stage of the point of absolute zero being passed was the fetus having come to be formed. The quantum level activity that took over was to develop that fetus within the *Hiraṇyagarbha*, the term literally meaning the womb bearing light, growing in space almost like a bulging womb and finally delivering what matured as a universe. In short, the so-called *kṣobha* was a mature universe reproductive impluse which initiated the conception of what was finally to appear as a universe. Almost along the line of biological reproduction, the contribution of Puruṣa, the efficient cause, was to the extent of creating a *kṣobha* within the inertial Prakṛti at rest in juxtaposition to it, and thence onwards it was all developed, delivered and maintained by *Prakṛti* as per the rules of the creation and behaviour, in short, the very nature of the infant universe that materialised. *Puruṣa* just remains a witness of the enitre behaviour of activity which runs as per conditioned by the mature universe cell or impulse called *kṣobha* caused by Its (Puruṣa's) instrumentality.

Within the *Hiraṇyagarbha* the transformation of *Prakṛti* to *ṛta* was confined while the *satya* started germinating in the form of sub-atomic particles both in matter and anti-matter forms. All of this has been named *ka* by Veda. It has been described earlier in Veda's own words through the elaborate interpretation of R.V. 10/121 in chapter 2. *Prakṛti* delivered the same in good time with a big bang in the form of immense amount of radiation and particles.

The illustration of a conception and later delivery of an infant universe must not be misunderstood with any physical contact coming to be established between *Puruṣa* and *Prakṛti* as the former has no physical existence and is utterly formless. It is the supreme dynamic wisdom in eternal omnipresent existence. Sāṅkhya, in an aphorism earlier than the one quoted above, suggests another simile.

'tatsannidhānād adhiṣṭhātṛtvam maṇivat' (S. 1/96). It says: as the magnet influences activity in iron so does activity get influenced in *Prakṛti* in the state of juxtaposition. This state of juxtaposition has already been discussed above. The point is that *Puruṣa* being incorporeal is not in and cannot establish physical or material contact with the *Prakṛti*, the material cause. But the former influences the latter in the said manner to conceive and deliver the universe at proper frequency.

Sāṅkhya has tried to remove a possible confusion by a reference from an Upaniṣad which may literally be interpreted to communicate the sense that *Puruṣa* created parts of the cosmos in a manner to mean that It happened to be the material cause. Sāṅkhya observation is: **'prakṛti vāstave ca puruṣasyādhyāsasiddhiḥ'**. It says: The influence of *Puruṣa* upon *Prakṛti* is in fact of a superintending nature. In other words the contribution of *Puruṣa* after generating the *kṣobha* in *Prakṛti* in order to initiate the conception of the universe, does not proceed in any manner in material sense except, so to say, keeping a watch or rather superintend the developments taking place at the level of *Prakṛti.* The developments are all material. *Puruṣa* as explained above, has already passed on or influenced the wisdom and the dynamism necessary for the work to be done along with the *kṣobha*. And as the creation and dissolution is an infinite affair, the constitutional limitations appear to operate simultaneously with the developmental process. But even though an army achieves the victory, yet the credit goes to the ruler; similarly in conventional terms *Puruṣa* is called the creator in spite of the creation having been physically done by *Prakṛti.*

One more quote from Sāṅkhya deserves consideration from the angle of *Puruṣa's* role in creation. It is: **"rāgavirāgayoryogaḥ sṛṣṭiḥ"** (S/2/9). The last word *sṛṣṭiḥ* means creation. In this quote Sāṅkhya defines as to what creation is? It says: The substantial attachment of *Prakṛti* associated with the detached dynamism and wisdom of *Puruṣa* is creation. This is to say that the entire creation called universe or any part thereof is the physical manifestation of *Prakṛti* or energy as called by modern science, but all the rules that govern it do so in a detachedly objective manner making no exception whatsoever in the consequences of their operation.

Attention is drawn to the part of the mantra (R.V. 10/129/1) mentioned above which describes the absence of both matter and anti-matter at the *mahāpralaya* or the state of great dissolution. Such a description also implies that the process of creation starts with the transformation of *ṛta* (K.E.) to produce matter as well as anti-matter particles which manifest simultaneously as described above. In chapter 8 it has been shown that the mention of matter waves denotes the birth of various particles etc.

Modern science supports such a position. The renowned scientist Stephen Hawking, in his book, *A brief History of Time*, says: "At the big bang itself, the universe is thought to have zero size, and so to have been infinitely hot. But as the universe expanded, the temperature of the radiation decreased. One second after the big bang, it would have fallen to about ten thousand million degrees. This is about a thousand times the temperature at the centre of the sun, but temperatures as high as this are reached in H-bomb explosions. At this time the universe would have contained mostly photons, electrons and neutrinos (extremely light particles that are affected only by the weak force and gravity) and their antiparticles, together with some protons and neutrons. As the universe continued to expand and the temperature to drop, the rate at which electron/anti-electron pairs were being produced in collisions would have fallen below the rate at which they were being destroyed by annihilation. So most of the electrons and antielectrons would have annihilated each other to produce more photons, leaving only a few electrons as left-over. The neutrinos and anti-neutrinos, however, would not have annihilated each other, because these particles interact with themselves and with other particles only very weakly. So they should still be around today. If we could observe them, it would provide a good text of this picture of very hot early stage of the universe. Unfortunately, their energies now-a-days would be too low for us to observe them directly. However, if neutrions are not massless, but have a small mass of their own, as suggested by an unconfirmed Russian experiment performed in 1981, we might be able to detect them indirectly: they could be a form of dark matter like that mentioned earlier, with sufficient gravitational attraction to stop the expansion of the universe and cause it to collapse again."

The first mantra related to the *Hiraṇyagarbha* appears once in R.V. and thrice in the Y.V., each time under a different *devatā*. The three appearances in Y.V. have been discussed in the chapter on Agni. The lone appearance is in R.V. 10/121/1. This has been discussed in chapter 2, its *devatā* or the subject matter is *ka*. The *ka* denotes a *Prajāpati*, *Brahman*, *growth*, *Agni*, *splendour*, *light* and *time*. The term *Prajāpati* in Veda denotes elementary state of creation; *Brahman* is that which expands; and light and time display its dimension. But to be more precise for scientific consideration all of these terms indicate the nature of the fluid state of Prakṛti or energy inside the bounds of the *Hiraṇyagarbha*. In short, the Vedic term *ka* denotes what may be understood as sort of primordial plasma in the modern scientific language.

The universe consists of manifestations which may be divided into three kinds : matter, anti-matter and black matter. Veda states the creation in a set of two terms: *sat* and *asat*. The *sat* denotes the world of matter which is characterised with multifaceted development. The *asat* is in fact non-*sat* and as such covers the world of the anti-matter.

Vedic Physics does not provide details in any length. One has to read within whatever has been said either in the Vedic text or by the ancient scholars in their works. In the chapter on *Prajāpatis* as the elementary creations, *Aṅgirā* has been clarified as denoting neutron among the elementary particles of Modern Physics and *Nirukta* has been quoted there in support of such an inference, which says that *Aṅgirās* were born of extremely hot burning stuff : *aṅgāreṣu aṅgirāḥ* (*saṁbabhūva*). This indicates the situation within the Vedic *Hiraṇyagarbha* or the primordial fireball of Modern Physics. The creation of hydrogen matter follows soon after the explosion when the neutrons in free space decay into protons and electrons, the latter being attracted to orbit the former forming hydrogen matter.

The initiation of the manifestation of *Hiraṇyagarbha* also effects the creation of *Ākāśa*, the field of motion and light. But all of these are rolled into *Hiraṇyagarbha* which is the singularity of space and time for Modern Physics. Kāla (time) is already marking time while the *Hiraṇyagarbha* comes into its own along with *Ākāśa*. The second *mahābhūta Vāyu* (force) also makes its appearance within the bounds of the *Hiraṇyagarbha*. Then comes the moment of the *sphoṭa* which modern science calls explosion. But the Vedic term is far more expressive and meaningful. The Vedic term for the developmental growth after delivery is *sphoṭa* as it means development or expansion. The Veda conveys that the so-called explosion was literally an expansion of the *Hiraṇyagarbha* itself into a universe. The apparent explosion itself is part of that process of creation.

The *mahat sphoṭa* or the big bang scattered the fireball in all directions causing the singularity to break into the expanding *Ākāśa* into directions and dimensions, radiating matter in the form of various *prajāpatis* as enumerated in chapter 6 and activating the time in creation.

As we shall see in the succeeding chapter on *Mahat Sphoṭa* which is a presentation of a whole 'sūkta' R.V. 1/163 briefly narrating as to what transpired as the galloping radiation, bursting forth from the *Hiraṇyagarbha* state spreading in and covering the vast expanses of *Ākāśa* (space) pervading all around, continued to cool down to form this universe in course of time. But before that it appears worthwhile to have a sketch of what Modern Physics at the beginning of the 21st century believes to have happened after the big bang.

Modern Physics believes that as the radiation flew in all directions after the big bang, the intensity of temperature fell rapidly. After about a hundred seconds it would have fallen to a billion degrees. This level of temperature is found inside the hottest stars. At this level the strong nuclear force would have become effective to produce deuterium nuclei by combining one neutron to one proton, which would have increased the combination to make Helium nuclei (two protons and two neutrons), and might even have succeeded in taking a couple of steps further along by way of producing small amounts of heavier elements. The free neutrons would have decayed to form the nuclei of Hydrogen atoms. The making of Helium and other atoms would have ceased within a few hours of the big bang while the universe would have continued to expand for the next million years or so with nothing much happening.

When the temperature would have come down to the extent that no nuclei and electrons were left with enough energy to overcome the electromagnetic attraction, they would have begun forming atoms. Such combinations would have created matter in heavier forms giving rise to comparatively denser regions within the universe that would have been expanding and cooling as a whole. The extra gravitational effect that would come into play would have effected a slow-down in expansion eventually causing a recollapse in regions where the motion would have dropped too far, to enable the gravitation of matter outside and around such regions to generate a slight rotational effect upon them. The collapse having commenced, such affected regions would have been becoming smaller in bulk—a situation that would cause them to rotate increasingly faster. As the region continued to be smaller and smaller, its rotation would be getting faster and faster. Finally a state would have been reached at which the speed of the rotation would have struck a balance with the gravitation countering the collapse. Thus the galaxies that rotate disclike came to be born. The regions which did not happen to be among those that had happened to pick up spinning would become elliptical galaxies because of their oval-shaped appearance. Such galaxies would not happen to be governed by an overall rotation, but their collapsing would have been countered because of their individual parts which would have established stable orbits around the centre of their respective galaxies.

Within the regions that would have taken the shape of galaxies the Hydrogen and Helium gas matter, with the passage of time, would have turned into smaller clouds which would tend to contract under their own gravity. As they became smaller in size their atoms would have tended to collide with one another resulting in a continuous increase in the temperature of the gas, which would have eventually started nuclear fusion. As a result of such fusion reactions the Hydrogen would get converted into more Helium, and the heat generated would cause the pressure to rise stopping further contraction of the cloud mass. Such blazing stable clouds would become stars, like our sun, burning their Hydrogen fuel into Helium and radiating light and heat all around.

The more massive ones, in order to counter their stronger gravity, would tend to be hotter to cause fusion reactions more rapidly, thereby consuming their Hydrogen in considerably shorter period which may be like a hundred million years. The conversion of Hydrogen into Helium would effect a contraction in their body, resulting in further rise in temperature within to start Helium burning and its conversion into elements like carbon and oxygen. And then, if the fuel lasts, with another contraction of the body the carbon blazing may commence the change to still higher elements. As described in the chapter on '*Pralaya*', with fuel lasting this process of burning and its conversion into heavier elements continues as far as iron where a continuously decreasing emission comes to an end and a crisis develops when the core of the star contracts to a state of neutron body or a black-hole, while the outer body explodes in a state of supernova outshining all other stars of the galaxy, and flinging the heavier elements into the galactic gas or dust thus providing raw material for the next generation of stars. Our sun is such a second or third generation star formed out of a cloud of rotating gas having the debris of earlier supernovas. Most of the gas of that cloud went into the formation of sun's body, but a few smaller gas clouds containing heavier elements contracted further to form the bodies of planets, one of them being our earth, which are now orbiting our sun.

In the following chapter, as mentioned earlier, a whole sūkta from Ṛgveda numbering 1/163 is being presented in full to show that it narrates the gist of what Modern Science is now finally saying regarding the sequence of events and developments of the birth and growth of our universe after having collected and verified the evidence with the instruments and resources that it has been able to invent and muster at its disposal to extend its eyes and ears to over a trillion galaxies spread over billions of light years in space and time with cameras and telescopes scanning and photographing details and even collecting the territorial material for lab analysis with the help of satellite and spacecrafts, at an unprecedented scale in human history. The said sūkta ends at the generation of stars followed by the advice to the mankind as to how the benefits with the employment of nuclear energy may be obtained by honouring the man of knowlege and science. The presentation of the 'sūkta' will be followed by a discussion of a Vedic noun for the sun which indicates that it is a second or third generation star born off the debris of on self-destroying supernova.

— 8 —

MAHAT SPHOṬA

Out of whatever models that have been advanced by modern physicists the Big Bang appears to have drawn fair amount of corroboration in subseqent researches. But it has not been able to meet the objections raised by quantum theory which seems to concede the phenomenon of the bang but holds that the initiation of the process of creation must have commenced prior to it and the bang occurred later whereby the initial growth got scattered in space. Reference of such a *sphoṭa* (bang) may be had in some of the sūktas of the 10th maṇḍala in R.V. but to our mind a concise description of what followed after the big bang is available in R.V. 1/163. The said *sūkta* is all the more important because the ṛṣi (seer of the said *sūkta*) states in one of its mantras as to how he could visualise that entire phenomenon. Towards the end, it has even been indicated as to how the nuclear energy may be obtained. That is why we have preferred to present the said *sūkta* in full in this chapter.

R.V. 1/163 begins with the bang having already occurred. So it does'nt describe the bang, but just mentions it. The *devatā* or the subject matter of the sūkta is *aśvogniḥ*. Translated into English it communicates the sense of a galloping radiation. That means that the *sūkta* presents a description of the radiation that got released with the explosion, the manner in which it spread and the forms the substantive part of it assumed. It should be kept in mind that Vedic terms are viable to be interpreted in spritual, intellectual and physical senses. As our subject is physics we would pursue the last track. It may also be pointed out that the way this event is presented by Modern Physics is more aptly described by the Sanskrit term *visphoṭa*' meaning an explosion, while our title above calls it a *sphoṭa*. The difference is revealed in the manner

Vedic Physics looks at the occurrence. The Sanskrit root *sphuṭ* means to develop. Hence its derivation *sphoṭa* means developmental growth. The prefix *vi* would just reverse the sense making it to denote destruction by explosion. According to Veda the said event was no destruction of what exploded. Instead the same rather blossomed to produce the universe.

Let us now begin consideration of the *sūkta mantra*-wise. We present the text followed by the meanings of Sanskrit terms, then a general translation and comments, if any, thereafter:

Ṛgveda sūkta 1/163:

yadakrandaḥ prathamaṁ jāyamāna udyan samudrāduta vā purīṣāt
śyenasya pakṣā hariṇasya bāhū upastutyaṁ mahi jātaṁ te arvan

(R.V. 1/163/1)

Term meaning: (yadakrandaḥ prathamaṁ) The way earlier roared; **(jāyamānaḥ)** that which was being born; **(udyan samudrāt)** rising from the ocean of energy; **(ut vā purīṣāt)** and/or from the debris of the previous creation; **(śyenasya pakṣā hariṇasya bāhū)** fast upon the wings of a hawk and the legs of a deer; **(arvan)** O Radiation! like a galloping horse! **(te)** your; **(ṃahi jātam)** great deed of emergence; **(upastutyam)** deserves being narrated.

Translation: O Radiation like a galloping horse! Being born, the way you first roared while appearing from the ocean of energy or the debris of the previous creation, fast upon the wings of a hawk and the legs of a deer, this great performance of yours deserves to be praised by way of narration.

Comments: As the Mahat burst forth with a roar, that is, a bang, the radiation appeared unbelievably fast from the ocean of energy or the debris of the previous creation, churning within. The Veda states an eternal cycle of creation and dissolution, hence the ocean of energy is equally well described as the debris of previous creation. The emergence was upon energy waves described like the wing motion of a hawk, and the matter waves described by way of the simile of a deer's run upon its legs. This wave description will be explained in greater details later. The *ṛṣi* says that such a happening deserves to be praised in words by way of narration in the following mantras.

yamena dattaṁ trita enamāyunagindra eṇaṁ prathamo adhyatiṣṭhat
gandharvo asya raśanāmagṛbhṇāt sūrādaśvam vasavo nirataṣṭṭa
(R.V.1/163/2)

Term meaning: (trita) floating superfast in space; **(yamena dattam)** released by the earlier controller; **(enam)** this radiation; **(āyunak)** got deployed in creative process; **(indraḥ)** electric charge; **(prathamo adhi atiṣṭhat)** had earlier already taken hold of; **(enam)** these radiated particles. **(gandharvaḥ asya raśanām agṛbhṇāt)** The space that held that galloping radiation caught hold of its rein, that is, checked its speed; **(sūrāt aśvam vasavaḥ nirataṣṭa)** the radiation that had emerged from the ball of fire, started decelerating in multiple locations.

Translation: Released in space the radiation in due course got deployed in creative process. The radiated particles were already electrically charged. The space slowed its high speed, i.e., went on cooling it effecting deceleration which caused emergence of denser and less denser locations.

Comments: The radiation, that got released from the Mahat spreading superfast floating in space, became deployed in the creative process. The particles radiating were already charged earlier negatively or positively as per the case. The mention of matter waves in the previous mantra is a clear pointer that the radiation bursting forth was full of matter particles as well. This is stating two aspects, the matter particles were charged and the radiation floating in space started getting slower and denser. In other words Veda is communicating that the intensely hot radiation went on losing temperature and its cooling down activated the process of creation.

asi yamo asyāditvo arvannasi trito guhyena vratena
asi somena samayā vipṛkta āhuste trīṇi divi bandhanāni
(R.V. 1/163/3)

Term meaning: (arvan) O galloping radiation! **(asi yamaḥ)** you are the enmass controller; **(asi ādityaḥ)** you are the sun; **(asi tritaḥ guhyena vratena)** you are endowed with the invisible characteristic of floating in space; **(asi viprktaḥ)** you have become divided; **(somena samayā)** due to cooling effect in course of time; **(āhuḥ te trīṇi divi bandhanāni)** It is said that you have three bounds in the field of light, that is, space.

Translation: (The last mantra has already stated the commencement of the process of creation.) This mantra says that the radiation addressed as arvan was to transform itself in bodies grouped enmass, the sun like stars, and all other bodies floating or moving in space on account of invisible characteristics. During the course of time it had so divided itself (to form such bodies) on account of the cooling effect within three bounds of space which are said to contain it.

Comments: Modern Physics states the details of what occurred immediately after the bang and soon thereafter when the matter created was at particle stage. The last mantra narrates the particle radiation from *Mahat* but skips the details. These have been summarily dealt with in sūkta R.V. 10/121 earlier in chapter 2. It proceeded to divide itself slowly in denser sections of substances that emerged in radiation and mass which was to get transformed into galaxies full of stars and their planetary worlds. The entire creation was to be confined within three bounds of space. The said bounds are length, breadth and altitude.

trīṇi ta āhurdivi bandhanāni trīṇyapsu trīṇyantaḥ samudre
uteva me varuṇaśchantsyarvan yatrā ta āhuḥ paramaṁ janitraṁ

(R.V. 1/163/4)

Term meaning: (trīṇi te āhuḥ divi bandhanāni) It is said that you have three bounds in the field of light, **(trīṇi apsu)** these three are also in particles/quantum state, **(trīṇi antaḥ samudre)** the very three are within the intervening space **(uteva me)** and similarly within me. **(Arvan)** O arvan ! **(varuṇaḥ chantsi)** you magnificent one are energising this performance; **(yatrā te āhuḥ paramaṁ janitraṁ)** right from where you are said to have had your excellent birth.

Translation: The repetition of the fourth part of the last mantra here as the first part is to lay stress upon the three bounds as the details of their realm follow with the statement that the three rule the particles/ quantum state, as well as the galactic and interspace among the bodies, and even the physical existence of the *ṛṣi*, the seer of this mantra. The field of light is *ākāśa* commonly understood as space in English and Modern Physics. Having emphasised that the three bounds are effective from the tiniest to the mightiest manifestations internally and externally both, the *ṛṣi* says that the magnificent radiation had been energising the process of creation right from the point of its birth onwards and triwards as it continued to cool fast. This implies the earlier formation of hydrogen and helium matter.

Comments: There were no axes prior to initial manifestation.

imā te vajinnavamārjanānīmā śaphānāṁ saniturnidhānā
atrā te bhadrā raśanā apaśyamṛtasya yā abhirakṣanti gopāḥ

(R.V. 1/163/5)

Term meaning: (vājin) A synonym for arvan; **(imā te avamārjanāni śaphānāṁ sanituḥ nidhānā)** these checks and brakes of yours applied for the safety of the bodies in fast motion; **(atrā te imā ṛtasya bhadrāḥ raśanāḥ apaśyam)** I saw the highly beneficial reins of yours here; **(yā abhirakṣanti gopāḥ)** which protect as invisible strings.

Translation: The *ṛṣi* says he saw the invisible force working like strings to keep these fast moving bodies reined for their own safety. These highly beneficial reins are creation of *ṛta* to act as checks and brakes upon these celestial bodies.

Comments: The *ṛṣi* is describing the role of *Vāyu* known as force to Modern Physics. The centripetal and centrifugal forces, in combination were and are in operation for the safety of each body in motion reining and keeping them out of collisions and along the safety of their respective tracks. These forces are themselves the creation of *ṛta*.

ātmānaṁ te manasārādajānāmavo divā patayantaṁ patangam
śiro apaśyaṁ pathibhiḥ sugebhirareṇubhirjehamānaṁ patatri

(R.V. 1/163/6)

Term meaning: (te ātmānam) Your eternally-in-motion form; **(divā patayantaṁ pataṅgam)** like the sun moving in the field of light; **(ārāt ajānām)** I came to know or realise from far and near; **(manasā)** with the instrumentality of my mind. **(apaśyam)** I saw or realised; **(avaḥ śiraḥ)** the all-savior or fore-part leading; **(jehamānaṁ patatri)** endeavouring like a flying bird; **(pathibhiḥ sugebhiḥ areṇubhiḥ)** along the paths easy to cover and devoid of any particles.

Translation: The *ṛṣi* realised that what had emerged as radiation has been creating a system in motion like the sun constantly moving in our near space. The system was the savior of the whole as stated in the previous *mantra*. Its leading part was covering, like a bird, areas of space that were hurdleless and easy to move in.

Comments: This *mantra* presents two points that need attention. It says **manasā ārāt ajānām** in the first part and **apaśyam** in the second part. The former communicates that the *ṛṣi* came to know or realise the whole phenomena from far and near with his mind. In the lower part he simply says that he saw. In Sanskrit seeing is also knowing or realising and *vice versa*. But he mentions his instrument, his mind, through which he came to know from far and near in the earlier part. This is a very clear indication of what we have mentioned elsewhere as well that he gained the knowledge whatever he is stating through the *mantra* by meditation in the state of *samādhi*. That is a state of extreme concentration—the psyche becomes oblivious of all save only the one it wants to know or see.

atrā te rūpamuttamamapaśyaṁ jigīṣamāṇamiṣa ā pade goḥ
yadā te marto anu bhogamānadādidgrasiṣṭha oṣadhīrajīgaḥ

(R.V. 1/163/7)

Term meaning: (yadā grasiṣṭhaḥ) Like a glutton; **(martaḥ)** a mortal; **(anu bhogam)** his choicest eats; **(ānat)** collects or possesses; **(iṣa ā)** to his heart's content; **(āt it)** similarly; **(atra)** here; **(te jigīṣamāṇam)** your wishing to be victorious; **(goḥ)** in motion; **(uttamaṁ rūpam)** best image or form; **(oṣadhīḥ ajīgaḥ)** obtaining burning material in immense quantity; **(apaśyam)** was seen by me.

Translation: The ṛṣi compares a big star which has amassed burning material in its body to a mortal glutton who stores up the choicest eats.

But while the glutton has been shown to be a mortal, emphasising the expectancy of an early death, the star is said to be desirous of being victorious in life, and collecting burning material with that purpose. That the *'ṛṣi'* has come to know or realise this state of affairs is apparent by his affirmation of having seen it.

Comments: The *mantra* discloses the scientific knowledge that the life of a star depends upon the quantity of its fuel, primarily the Hydrogen matter which may last long. A star would survive as long as its fuel would last initially in the form of Hydrogen which gets converted into Helium with its temperature having risen to Helium burning point thereby converting the latter to Carbon and so on. The Carbon burning requires temperature rise enough to cause its flash point, finally a supernova leaving behind a neutron body advancing to become a black hole. It is remarkable to note that the term used for the matter to be stored in its body by a star at the initial stage does not state the name of Hydrogen. It says **oṣadhīḥ**. This term derived from the root *us* to burn denotes atomic fuel, covering subsequent atomic fuel formations as well. This atomic fuel is in fact the choicest eat for a star, and the term atomic fuel is rightly in vogue with the modern physicists.

anu tvā ratho anu maryo arvannanu gāvo' nu bhagaḥ kanīnām
anu vrātāsastava sakhyamīyuranu devā mamire vīryaṁ te

(R.V. 1/163/8)

Term meaning: (anu tvā rathaḥ) In accordance with your process of creation; **(anu maryaḥ)** in accordance with the young stars; **(anu gāvaḥ)** in accordance with the planetary systems; **(anu bhagaḥ kanīnām)** in accordance with those expected to be born off the maiden stars; **(anu vrātāsaḥ)** in accordance with the code of the manifestation of satya; **(arvan)** O one appearing as galloping radiation! **(devāḥ tava sakhyam īyu)** the radiant manifestations i.e., stars, etc., are desirous of your friendship; **(mamire vīryam te)** and will manage to pursue the process of creation in the steps of your valour.

Translation: In accordance with the process of creation pursued by the galloping radiation, in accordance with the way new-born stars come up, in accordance with the manifestation of the planetary systems, in accordance with the generation of stars expected to be born, in accordance with the code followed by the manifestation of *satya*, the system so far created and to be further created, will remain desirous of friendship to whatever has been created and the manner in which the same has been created, the radiant stellar systems would continue creation in the same footstepsso declares the *ṛṣi*.

Comments: According to the *Vedic Koṣa*, a dictionary of Vedic terms as interpreted by Maharshi Dayānanda, **rathaḥ** means *vijñānam*, that is, the science of creation in the present context. Our reference dictionary states **maryaḥ** to mean a young man, as such the same has been taken as a young star here. The term **bhagaḥ** is used for the sun yet to rise and **kanīnām** denotes the plural form in relative case of *kani* meaning a maiden, hence is interpreted as stars to be born. **Vrātāsaḥ** as per the *Vedic Koṣa* referred to above, as **satyācāra** is the code of the manifestation of *satya* as the cosmos is but a manifestation of satya, i.e., matter. The verb **mamire** is derivation from root **mā** to mete, to take or provide form, which the **devāḥ** in mantra are obliged to do following the steps of their creator. Veda has also narrated generations of the stars that are still to follow as in the state of being under description.

hiraṇyaśṛṅgo ayo asya pādā manojavā avara indra āsīt
devā idasya haviradyamāyan yo arvantam prathamo
adhyatiṣṭhat

(R.V. 1/163/9)

Term meaning: (hiraṇyaśṛṅgaḥ) The peak of illumination; **(ayaḥ asya pādāḥ)** its mobile body having reached iron stage; **(manojavā)** superfast/extremely fast; **(avara indra āsīt)** resplendent like a second sun; **(yaḥ arvantam prathamaḥ adhi atiṣṭhat)** which rode upon the galloping radiation earlier; **(asya haviradyan)** whatever was eaten by it; the same **(devāḥ)** other celestial bodies; **(āyan)** did receive for consumption.

Translation: The *mantra* is describing a star which gets so intensely illuminated that the same has been described as the peak of brightness; it has reached iron stage in itself; its light is shooting superfast; it appears brilliant like a second sun. It had ridden upon the galloping radiation earlier, that is, it got manifested earlier by the said radiation. The matter that was in its body is to be consumed by other celestial bodies.

Comments: The *ṛṣi* is describing a supernova. Apart from other characteristics the mention of **ayaḥ**, that is, iron is most pertinent scientifically. It unmistakably indicates the end of carbon burning. The brilliance of its burst has been described as if a second sun has appeared in the sky, an event visible from the earth. But the mention of **ayaḥ**, iron, as its **pādāḥ**, reach, is a feature which either Modern Physics knows or the Veda, the oldest book in human library, has been telling this to the world. The term **pādāḥ** is disclosing that the superfast activity generated due to the burst or explosion is the result of the internal carbon burning having reached the iron stage and that the atomic fusion within its interior is no more remitting any energy. Through a supernova explosion the star throws off most of its body matter into space around which either gets incorporated into the bodies of other stars under gravitational attraction, or becomes part of any new star, star dust or planet of the following generation on account of the same. This is what has been so well stated in the latter half of the *mantra*. The Vedic name **hiraṇyaśṛṅgaḥ** for the modern supernova is no less meritorious.

irmāntāsaḥ silikamadhyamāsaḥ saṁ śūraṇāso divyāso atyāḥ
haṁsā iva śreṇiśo yatante yadākṣiṣurdivyamajmamaśvāḥ
(R̥.V. 1/163/10)

Term meaning: (irmāntāsaḥ) Such long distance stars whose rays end up trembling at earth; **(silikamadhyamāsaḥ)** such systems which are centrally controlled; **(śūraṇāsaḥ)** star systems arrayed in columns or other formations; **(divyāsaḥ atyāḥ)** the heavenly bodies in constant motion; **(aśvāḥ)** these fast moving bodies; **(haṁśāḥ iva śreṇiśaḥ yatante)** are endeavouring like swans flying in formations; **(yat)** so that; **(ākṣiṣuḥ)** they continue to obtain/remain upon; **(divyam ajmam)** their divine paths/tracks.

Translation: The mantra says that all of the heavenly bodies in motion including such whose rays end up trembling upon earth or those belonging to systems like our solar system or those moving in a galactic formation, or all of those which are constantly in motion in the heavens, they are on the move like the swans flying in formations endeavouring to stick to their respective tracks in space.

Comments: The Veda states the settled state of creation in space. The *mantra* sums up the various heavenly bodies in its fore-part and reports the same with a simile with swans which always fly in formation. What is interesting about this simile is that swans fly out of their nests in formation and return to their nests likewise. Veda implies that all these kinds of heavenly bodies are endeavouring to keep track of their respective paths in this long journey from creation to dissolution. There is never a question of any disorder. They have risen in an order and would collapse in an order. Each body is riding a track and all tracks are determined by the cosmic order.

Another point noteworthy from scientific angle is the use of the term **īrmāntāsaḥ** which denotes the end-trembling rays. The Veda does not state that the rays coming from the stars are trembling. The term specifies their end-trembling characteristic. We know that the rays do not tremble. They just appear to tremble due to earth's atmosphere at the end of their journey upto this planet.

tava śarīraṁ patayiṣṇvarvan tava cittaṁ vāta iva dhrajīmān
tava śṛṅgāṇi viṣṭhitā purutrāraṇyeṣu jarbhurāṇā caranti

(R.V. 1/163/11)

Term meaning: (tava śarīram) your physical body; **(patayiṣṇu)** is desirous of motion; **(tava cittam)** but your instinctive intelligence; **(dhrajīmān)** is agile; **(vāta iva)** like the wind. **(tava śṛṅgāṇi)** your *tejas*, that is, light/heat rays all over, **(purutrā jarbhurāṇā)** are intensely active in many ways; **(araṇyeṣu caranti)** and are coursing through the barren parts of inner and inter space.

Translation: The Veda states that matter is liable to change. (The next *mantra* would disclose the method to effect beneficial changes in matter.) But behind such a changing character of the body of matter is the agile instinctive intelligence imbued at the state of Mahat. The later half of the mantra sums up by pointing out that the universal system, already created, in its own way, is active in barren areas, implying improvement in conditions there in course of time.

Comments: The Veda changes gear. The idea is that the crown jewel of the cosmic reproduction is the emergence of highly intelligent species like human beings. The system, in its own way, and in its own time, is to create habitable territories, like this earth, by way of growth and development of physical conditions with suitable atmosphere, etc.

upaprāgācchasanaṁ vājyarvā devadrīcā manasā dīdhyānaḥ
ajaḥ puro nīyate nābhirasyānu paścāt kavayo yanti rebhāḥ
(R.V. 1/163/12)

Term meaning: (devadrīcā) By honouring the men of knowledge and wisdom; **(manasā)** through the application of mental power **(dīdhyānaḥ vājī arvā śasanam)** the method of magnificent superfast radiating energy for effecting transformation by way of destruction **(upaprāgāt)** be learnt from nearby; **(ajaḥ puro nīyate nābhiḥ)**; the birthless leads to the nucleus from the fore; **(rebhāḥ kavayaḥ)** brilliant persons steeped in the science of physics; **(yanti)** endeavour the same **(asya anu paścāt)** by following the process later in reverse as well.

Translation: The *mantra* states that the thermonuclear power employed by the magnificent superfast radiation may well be obtained from nearby by honouring men of science who may do it by applying their mental prowess. The birthless (energy) leads to the nucleus from the fore, but the brilliant persons well-versed in the science of physics can follow the same method of approach from the fore or they may also obtain it through reverse approach later.

Comments: Hardly any after such a forthright statement by Veda. Little wonder that the West discovered nuclear power after Vedas had reached there and they had been dissected by brilliant minds. The mode of approaching the nucleus from fore is the principal method in the process of creation employed in the universe wherein each star is a virtual nuclear furnace. The reverse method of approaching from behind was the first to be discovered and employed by the Americans when they nuclear-bombed Hiroshima and Nagasaki in Japan during the World War Second. In scientific terms the two are respectively known as the fusion and the fission methods.

upa prāgāt paramaṁ yat sadhastham arvāṁ acchā pitaraṁ mātaraṁ ca
adyā devāñjuṣṭatamo hi gamyā athā śāste dāśuṣe vāryāṇi

(R.V. 1/163 completed)

Term meaning: (yat) That; **(juṣṭatamaḥ acchā arvā)** superbly well wishing body of wise men; **(dāśuṣe hi)** for definitely providing; **(atha)** time and again; **(adyāḥ gamyāḥ vāryāṇi)** utilities, travel facilities and other desirables; **(devān śāste)** may seek control of divine capabilities; **(upa prāgāt)** by gaining or obtaining; **(pitaram)** a protector; **(paramaṁ sadhasthaṁ)** top-grade nuclear reactor plant; **(mātaraṁ ca)** besides its full knowledge.

Translation: The *mantra* in continuity of the preceding statement about the process of obtaining thermonuclear power, remarks that a superbly well wishing body of wise men should have a top-grade atomic plant as a protector, besides its full knowledge, to provide time and again power for various utilities, travel facilities and other desirables or contingencies, by seeking control over the capabilities inherent within creative process. The expression **devān śāste** communicates the sense of disciplining the powers of atom.

Comments: The Veda while introducing the power that created the universe further states that the same may be harnessed for improving the standard of life but warns that its source must be kept under the control of a superbly well wishing body of wise men (read scientists). The word *pitaram* in the *mantra* above having been derived from root *pā* to protect, has been translated as a protector; *sadhastham* literally meaning an associate placement, has been interpreted as a virtual steller furnace on earth, which is interpreted as an atomic reactor in the present context. Similarly '*mātaram*' has been stated as skill or knowledge of generation.

Before ending this chapter dealing with creation of our universe as stated by Veda, let us as well consider a name given to our sun by Veda which reveals the story of its birth. A Vedic name of our sun is *mārtaṇḍa*. Literally it means born of a *mṛta+aṇḍa* (dead egg). But let us analyse it with the Vedic technique. *Mṛta* is dead alright, but *aṇḍa* is derived from root *am* to yoke together with suffix *ḍa* under *auṇādika* rule 1/114 thereby meaning driven by two yoked together. In the universe all of the celestial bodies are being driven by the two forces, centripetal and centrifugal, yoked together, and as the stars are the ones which are capable of producing other bodies, the body under consideration ought to be a star. In the light of this analysis the use of such a noun for our sun denotes that the same is born of the debris of one or more than one dead stars. The Vedic Sūkta R.V. 1/163 presented and explained above, in its 9th *mantra* describes only one star which dies so brightly as a second sun and blasts its body matter into space generating intense heat and light many times that of our sun, thereby causing production of heavier elements. Modern science calls it a supernova. It has further come to be learnt that our sun is born of the debris of one or may be more than one supernovas. Veda had long ago revealed such a birth of our sun and continues to do so by naming it *mārtaṇḍa*.

It may be added that this description by Veda is very remarkable as Modern Physics is trying to arrange its knowledge regarding the origin of creation along the same line. That is why we have preferred to quote the original Vedic text and translated the same termwise followed by our comments. How could Veda know is a question to be answered and deliberated upon by the scientific community of our times. The Veda and the seers of its *mantras* have their answer equally emphatically written in the Vedic text itself. All they expect is a little more reverential approach.

— 9 —

TANMĀTRA

Sāṅkhya at 1/61 after *Mahat* and *Ahaṅkāra* proceeds further by stating **"ahaṅkārāt pañca tanmātrāṇi"**, that is, from *ahaṅkāra* emerged five *tanmātras*. What is this *tanmātra*?

In the chapter on *Mahat* earlier it has been pointed out that the *kṣobha* or agitation in *Prakṛti* (energy) at rest state, effected the rise of temperature under sub-absolute zero state. There is no way to know how much below the absolute zero point it was because the state is beyond measurement or even calculation. All that can be said about it, is that it was an extremely fine state of existence. It was totally a no-gross state. *Prakṛti* itself being inertial was incapable of making any slightest stir. It just existed and was at rest. Then, what was the source of that agitation?

Modern science has no answer. But Veda declares, and on its authority a whole host of most ancient literature produced by the seers of Veda, affirms that it was *Puruṣa*, the infinite totality of dynamic wisdom, which enjoys existence pervading and transcending *Prakṛti*, that was instrumental in causing the said agitation. Its will is effective even in a state of existence well beyond the realm of relativity, a state which was yet to be created. The said will causing impact in a state of measureless existence, was itself measureless or *amātra*, that is having no *mātrā* or measure.

As rise in *tapas* or temperature began it started assuming some proportion, *mātrā*, which became measurable soon after it passed the absolute zero temperature point in modern scientific terminology. It had assumed what we call today quantum size. Thence each impulse that got added to its proportions was a multiple of that quantum measure.

Modern science has moved from the macro to the micro. It first discovered the field of relativity and later on found out that the energy works through the quantum multiples and it is such a working which assumes mega proportions. It is something like seeing the cart first and its horse later. But having reached the final conclusion that the process of creation moved from the micro to the macro, at least it ought to be presented the right way. Almost in all the textbooks of Physics we first meet Einstein and then come to Max Planck and others who have reported their quantum findings in various fields. But Sāṅkhya proceeds from *Prakṛti* and presents its manifestations in the order they occur. From *Ahaṅkāra*, which has been stated as the cause of *tanmātra*, a noun displaying a fundamental characteristic covering the entire manifestation, Sāṅkhya proceeds to enumerate the causatives of the matter, and *tanmātra* state being the precursor of this relativity, is being detailed first.

The term *tanmātra* is a compound of two words: '*tasya + mātram= tat + mātram = tanmātram*'. The word *mātram* in Sanskrit denotes smallest unit of measure. Hence this compound word denotes it as that little unit or quantity; the pronoun is indicating the related subject. Or we may say merely that little. Modern Physics is using the term quantum in this respect. But the word quantum denotes the sense of how much. The Vedic term *tanmātra* is a positive statement of the state of being.

Moreover by enumerating *tanmātram* or five *tanmatrāṇi*, the latter being a plural of the former, Sāṅkhya appears to state, that *Prakṛti* or energy is quantized—a position Modern Physics holds today. But on second thoughts the actual state of affairs appears to be slightly different. Vedas, it appears, do not say anything special about the state of being of *Prakṛti*. Whatever thcy say has been well summed up by Sāṅkhya in its opening statement at 1/61 by stating that it remains conserved under any of its three attributes. *Prakṛti* is held as an *upādāna kāraṇa,* the material cause of creation, which being inertial, is unable to take any step on its own. And Sāṅkhya is stating five kinds of *tanmātrāṇi*. It would be presumptuous to conclude that *Prakṛti* is quantized in five ways.

Veda and Sāṅkhya both lead us to five mega causatives to emerge from *tanmātrāṇi*. So far as the induced activity of *Prakṛti* is concerned—we call it induced because by itself it is incapable even to stir, it ought to be on account of *Puruṣa* and the Sāṅkhya declares that the activity occurs at *tanmātra* level. The present physicist also has no proof of the energy being quantized save the fact that it has been found to act at quantum level, and being inertial it can hardly behave in that manner unless it exists quantized. As we shall see a little later, modern science is still busy in getting acquainted with more and more ramifications and reactions at micro level of different aspects of the creative process. They will be taken up a little later in this very chapter. So far as Vedas are concerned, they cannot change while the science, with each remarkable discovery, has been changing, nay, correcting its stance, of course with the loftiest goal of determining the reality.

Whatever be the state of being of *Prakṛti* or energy at rest, it may assuredly be concluded that its emergence as the material creator of universe gets initiated at quantum level. This leads to the number of *tanmātra* which is five. It is because Sāṅkhya maintains that there are five different mega causatives which contribute to finalise the creation of matter like working an assembly line. The character of each of their respective contribution is dissimilar to the other and yet complementary towards the end purpose. At each stage the quantum unit appears in a different garb. That is why Sāṅkhya speaks of five *tanmātrāṇi*.

But the word *quantum* is in use in modern science in a very limited sense exclusively, that is, to denote the quantum of energy released by an excited electron orbiting a nucleus. It is better known as a quantum of electromagnetic energy. It is not in use to denote the minimum quantity of force or heat/temperature or such minimum units causing interactions at respective stages or levels of activity. Probably the reason is that such discoveries have been made from time to time at different intervals in different spheres of activity and each one got named like a new-born baby.

Vedas specify the process of manifestation in five separate heads of causatives, each making a distinct contribution not common with rest of the four. These causatives are *Ākāśa*, *Vāyu*, *Agni*, *Jala* and *Pṛthvī*.

The effect caused by each one of them is complementary in nature to transform energy into matter after the final interaction gets, so to say, executed by the fifth causative in succession. Detailed presentation of these causatives, which together cover the entire range of relativity, will be taken up in the following chapters. This mention is to clarify and explain the five kinds of *tanmātra* specified by Sāṅkhya as emerging from *ahaṅkāra* so that they could easily be so grouped. The compound *tanmātrāṇi* may easily be adopted with a slight change like *ākāśa mātram*, *Vāyu mātram*, etc., to denote the nature of specific interactions.

The entirc universe is in space. That is where all manifestations have taken place, do take place, and shall continue to take place as long as the creative process is on. All events occur within the same, and each manifestation, right from the micro to macro, does occupy an extent of space. What is the smallest unit of this space ? The Vedic term that represents space is *ākāśa*. It means the field of light. Naturally the particle of light is the minutest unit, to occupy and represent space. It is the quantum of electromagnetic energy, called photon. That is the first amongst the five *tanmātras* stated by Sāṅkhya and may well be named *ākāśa-mātram* which is the cause of the creation of *ākāśa*, the field of relativity.

The second causative is *Vāyu* which, as shall be explained at length later, is known to modern science as force which is recognised in four states, namely gravitational, electromagnetic, strong nuclear and weak nuclear forces. The term *Vāyu* covers specifically four types of interactions, namely, *sūcana*, *utsāha*, *hiṁsana* and *prakāśana* which respectively and justly cover each of the four states of the force specified by modern science as mentioned above. This interpretation has been discussed and presented as supported by the quote from the lexicon stating such a meaning of Vāyu in a chapter under that title later. As such the graviton of gravity, the special photon which keeps on exchanging during the electromagnetic activity, the pion which keeps on ball like exchanging position during the working of the strong force within the nucleus, and the so-called W particle named after the first letter of the weak force may conveniently be recounted as *sūcana mātram*, *utsāha mātram hiṁsana mātram*, and *prakāśana mātram* or even just *pra-mātram* in the fashion of W particle respectively.

The third mega causative is *Agni*, which term covers light, heat, temperature, etc., universally. The three kinds of *Agni* include nuclear, electrical and physical states, so specified by Veda. Modern science has developed different terms to name the measure in each case. The unit to measure heat is calory, for temperature it is degree and for the amount of electrical energy that changes to heat or light it is joule. So far as the temperature is concerned there are three kinds of thermometers in vogue. The measure of degree is virtually the same in Kelvin and Celsius thermometers, the difference between the two being the placement of a 0 degree mark. What is 0 in Kelvin is –273 on Celcius. The Fahrenheit, also named after its maker, has a different measure of degrees. The equivalent for heat is *tāpa* and temperature is *ūṣmā* in Sanskrit. The calory may well be represented by *tāpa-mātram* and the temperature as *ūṣmā-mātram*/*ūṣmāṁśa* as aṁśa is degree. For absolute zero *parama śūnya* is already in vogue. So far as joule is concerned it being a transfer measure may be termed as *madhya-mātram*.

Next comes *Jala* the fourth among the causatives. This is the most vital stage in the manifestation process. According to Veda all transformation is ***agnisomābhyām***, due to *agni* and *soma* acting as complementary contributing factors. *Agni* represents the heating effect and *soma* works as the cooling one. All objects in universe from micro to macro, are products and bearers of certain amount of intensity of heat which has stabilised within it at that level, and thus retains it in that state. *Agni* cannot cool itself. It is '*Jala*' which marginalises *Agni* at each rung of the atomic manifestation, either in the form of isotopes or atoms. All fusions take place at certain temperatures, and *Jala* stabilizes them at that stage.

The process begins at hydrogen level. A hydrogen atom is the simplest among the family of atoms, and the first step. It has a proton in its nucleus and an electron orbiting the same. It is a stable atom. The rising heat factor effects the entry of a neutron and *Jala* stabilizes its fusion at certain temperature. Such a fusion is effected in two manners, either as a neutron within the nucleus, or in the two parts of a decayed neutron, i.e., a proton and an electron, in which case the proton gets added to the nucleus while the electron goes into its orbit. It is *Jala* that determines that at each step of atomic growth the additional mass involved in fusion is neither less than, nor in excess of, a neutron mass, either as a neutron or as a proton and an electron. The quantity in excess is thrown out and the rest is stabilised as an isotope or an atom. An isotope results with the fusion of a neutron, while atom gets stabilized when an electron gets added to the orbiting team, and proton to the nucleus. Whatever be the mix-up in the cauldron, the final stabilisation gets effected in the manner indicated above,—the decisive mātram unit is the electron which effects change in the chemistry of the element by joining its orbit or simply makes addition to its weight when entering the nucleus as a part of the neutron mass.

It is again the electron which reacts as a *tanmātra* in effecting a bond to help the formation of a molecule under the influence of the fifth and the final causative, *Pṛthvī*. Depending upon its reaction as stated above the electron behaves as *jala mātram* or *pṛthvī mātram.*

Vedic Physics emphatically states that all macro size developments within the range of relativity get initiated at micro level which forms its *mātram.* As micro leads to macro, we may as well turn to the five *mahābhūtas* (causatives) which follow one by one after an introductory chapter. In other words five different kinds of interactions are involved in the manifestation of matter, hence quantum activity begins at five micro levels each growing to macro state in a sequence, the former called *tanmātra* and the latter *mahābhūta* by Veda.

— 10 —

PAÑCA MAHĀBHŪTA

In the previous chapter it has been explained that according to Veda and Sāṅkhya treatise *Prakṛti* initiates its process of transformation into matter at minutest level including the elementary particles. Such a transformation process has been stated to occur in five stages depending upon the five different kinds of behavioural patterns displayed at the level of initiation. As each pattern of such activity grows to cosmic proportions they are called *pañca mahābhūta* or the five mega elements each of which originates in succession from its precursor. They are named as (1) *Ākāśa*, (2) *Vāyu*, (3) *Agni*, (4) *Jala* and (5) *Pṛthvī*. The minutest beginning process works in different forms in each causing the initiated micro process of creation to assume mega proportions.

Modern Physics has now discarded the four fundamental elements of creation the ancient Greeks believed in, namely, air, fire, water and soil. They were adopted as such by the Christian theology and thus came to assume a divine character. At the time of the advent of Islam these four also came to be adopted, as they were, in the Islamic theology almost in the same context as above. Thus in course of time with the spread of Christianity and Islam in various parts of our world, the four came to be believed as the fundamental elements of creation till such time the scientific researchers in the west girded up their loins to demolish them even at the cost of personal pain. The restraining hand of the Church is still in evidence although comparatively ineffective now.

In India, the land of the revelation of Veda itself, and the composition of the six Vedic treatises including Sāṅkhya, on account of various upheavals, a communication gap distanced the people from the long established system of Vedic studies resulting in illiteracy and loss of scientific temper. The political subjugation for centuries helped the degeneration of character, wisdom and intellectual acumen. On top of it the double century British rule left no stone unturned with a view to rear a colony of slaves to work for the growth of their empire. Their scholars studied in depth the Sanskrit books and realised that any indepth support to Sanskrit education would work for their own defeat. The study of Sanskrit along the age-long established system is known to develop scientific temper in routine course.

Under conditions briefly narrated above the scientific knowledge inherent in Vedic terms was forgotten and the conventional meanings assumed a dominant role in seeking interpretations. According to the conventional meanings of the five *mahābhūtas* (mega causatives) named above, *Ākāśa* came to mean sky, Vāyu air, Agni fire, Jala water and Pṛthvi soil, or the earth. It is rather interesting to note that the four fundamental elements enumerated by the ancient Greeks, the Christian theology and the Islam virtually aver to the last four of the five mega causatives doing rounds in India. How is it that these four with an established mark of Indian origin more ancient than the age of Aristotle and Plato in Greece, came to be so very well accepted by the later, from whom they are believed to have been adopted by the Christian and the Islamic worlds? The only valid reason that holds water appears to be that they spread out of India both westwards as well as eastwards much before the advent of Jesus Christ in the land of Yahweh. The Greek scholars, who were men of no mean intelligence, pondered over them, and believing the last four to be making a material contribution to the existing matter and life available upon their home planet, and finding no such physical contribution of sky, the latter four were assimilated in their own explanation of the phenomena of creation while the first one got discarded. Christianity and Islam, both having no theory of their own, came to accept them in normal course.

But the so called four basic elements, namely, air, fire, water and soil, are no fundamentals as they themselves are products of other elements. As the physicists took to experimentation, they found out the truth and the truth prevailed. Christian theology conceded the dawning of scientific knowledge reluctantly and Islam is bound to follow suit sooner or later as the Muslim physicists have given the truth a *de facto* recognition.

We observe that universe is located in three dimensional space all around. All of its activities are taking place in this space itself. The world around us, the planets, our sun, and the multitude of stars shining our nights, exist and move within the expanse of space. The term *space* is the name accepted for this vastness by Modern Physics. Veda calls it *Ākāśa*. The Vedic as well as the English terms, namely *Ākāśa* and *space* will be discussed at some length in the next chapter.

According to Modern Physics, at the time of commencement of the creation of universe, the first to come to manifest was force. It appeared at the dawn of creation in space already in existence. That *Ākāśa* gets created as the first mega element to put in appearance is a Vedic statement. No such claim is made by the physicists in respect of space. Force is responsible for the maintenance of order in the cosmos. Its control extends to each one of the objects of creation big or small. Sir Issac Newton has called force to be an external agent exerting influence upon the physical bodies. Veda traces its origin to the inspiration of eternally existing supra dynamism, which also effected the arousal of *rajas* (motion) in dormant *Prakṛti* and transformed the same into K.E.. The superb order in universe owes its credit to this force.

It is force which keeps an electron in orbit around a proton in a hydrogen atom and the rotating nucleons of isotopes and atoms to the confines of their nuclei with electrons doing the rounds in orbit. It is force which unifies an electron to a proton transforming the two to a new particle called neutron. On a bigger scale it is force which keeps the moon rotating and running with the earth, the planets, including the earth, doing the same drill with the sun, and the sun, in its turn, marching in a similar fashion, with its entire family in tow, along its orbital course, as one of the stars in the Milky Way. Nothing within the physical limits of creation is beyond the influence of force.

This force comes as the second *mahābhūta* mega causative after the first, *Ākāśa*. Veda and Sāṅkhya name it as *Vāyu*. It is again specifically stated that though the conventional or traditional meaning of *Vāyu* is air or wind, yet in this particular Vedic context it denotes force. The etymological analysis will be presented in an independent chapter on *Vāyu* which follows later. This *Vāyu* originates from the first *mahābhūta* and works in succession to the same.

The science of physics tells us that Hydrogen atom, the first on periodic table, and the only one based on a combination of two particles, an electron and a proton, is a stable atom. This means that the binding electro-magnetic force between the two particles is rather strong to hold the two together in a stable condition. As the creation of universe can take place only if the mass of the hydrogen matter created at the initial stage, gets transformed into other atoms, molecules and compounds, etc., the creative process must proceed by destabilising the hydrogen stability. This is achieved by the third *mahābhūta*, Agni, which originates from its precursor *Vāyu*, and works as its successor.

Agni is not the ordinary fire as it is understood to mean conventionally. It is in fact the thermo-nuclear fire, that is the fire due to the heat caused to be created within the nucleus under gravitational pressure causing friction. We shall see in the chapter on *Agni* later how such a heat gets generated and creates conditions conducive to its job. The rising temperature reaches stages for the fusion of a neutron mass. A neutron is a sub-atomic neutral particle. It remains stable when it is a nucleon, i.e., in a state of being within the nucleus of an atom. A free neutron in a period of approximately 12 minutes decays into an electron and a proton.

The fourth *mahābhūta Jala* causes the growth of hydrogen atom either by way of the addition of a neutron to nucleus thereby effecting a physical change in the atom in its new form of an isotope or by fusing a proton within the nucleus and an electron in its orbit thereby bringing about both a physical as well as a chemical change by way of producing a new atom. Again it is pointed out that the noun *Jala* does not denote its conventional meaning, i.e., water, as is generally understood. The etymological meaning applicable in Veda will be explained in a later chapter devoted to it exclusively.

We find that the process of effecting growth of an atom, either by way of assimilation or rejection or both, gets completed after the action of the fourth *mahābhūta Jala*. But the state of stability of energy in the new form still remains wanting. This is achieved by the fifth and final *mahābhūta* called *Pṛthivī*, which ties the atoms together in the form of molecules, with the creation of bonds in between them. Such an action brings about expansion and thereby stability in the atomic material. Various bonds get created resulting in numerous forms of molecules which are the building material of universe. It may be mentioned that *mahābhūta Pṛthivī* puts in its appearance as a follow up of Jala, and that its name must not be confused with any conventional or traditional meaning attached to it. This aspect will be explained fully when Pṛthvī is discussed in greater details in a separate chapter to follow.

The process of transformation with which energy changes itself to matter is very well known to Modern Physics. The same is so well known that the leading physicists have successfully employed energy in their labs by developing such mechanical set-up and devices which can not only duplicate the conditions prevailing in nature for the creation of matter, but have even succeeded in producing atoms now being shown on the periodic table from No. 93 onwards to well beyond the 100 mark, all of them being endoergic, and not available in nature. What we want to point out is that nowhere Modern Physics has been able to discover something that oversteps the delineation of the fields of interactions by Veda.

On the contrary Vedas have been presenting knowledge regarding the process of creation couched in terms which are self-explanatory in revealing scientific knowledge which the modern physicists have been able to discover only recently, and are still busy in trying to understand certain aspects either more positively and/or more fully. The reason as to why Sāṅkhya, while enumerating these causatives of the creation of matter, called the five as mega elements which follow the five *tanmātras*, to our mind is that Vedic seers had realised the total import of such a knowledge from the sole source of existence that energised inertial *Prakṛti* itself to all of this purposeful activity.

As enumerated specifically by Sāṅkhya, the physics as stated by Veda presents the process of creation or transformation of energy into matter resulting in manifestation of universe as a series of four separate inter-actions initiated at the minutest levels which expand to cosmic proportions. Four of these interactions occur consecutively in the field of light which happens to be the first requirement of creation. It is *Ākāśa* which is the first *mahābhūta* and the rest to effect respective interactions are the remaining four *mahābhūtas* that occur in the order of sequence as mentioned. All along the electron leads the way. It is like the functioning of the universal automatic plant of *Prakṛti* being described. A plant requires space or field where it can function. There a four-stage assembly line is so arranged that at the end each stage feeds its product to the next. The plant feeds upon the energy and the product is matter. Well, not exactly, but this is merely a simile and all similes are partial. In such an imaginary plant, the energy is the feed, worker, machinery and the product itself.

Before we take a closer look at such a set-up, one point needs to be clarified. What we have been calling *mahābhūta* have been named *sthūlabhūta* by Sāṅkhya. The term bhūta means that which has manifested, and *mahā* means mega. Hence the five mega elements of creation. But the adjective *sthūla* employed by Sāṅkhya traditionally means fat. To understand what the Sāṅkhya is denoting the derivational approach must be taken. Derived from the root *sthūl*, to increase, the term means that which has increased in size or scale. The *sthūlabhūta* is the increased state from the minutest level. To avoid confusion we have used the term *mahābhūta* which is synonymous to Sāṅkhya term and all of their work on cosmic level.

— 11 —

ĀKĀŚA

The creation of universe requires a field where it may be located to carry on its expansion as per the laws of nature. According to Modern Physics universe is located in space which extends in all directions. Veda names it *Ākāśa* which is described as the *adhiṣṭhāna* (substratum) for the same. Let us consider these two names which are considered synonyms in a conventional sense.

Space is defined in the Random House Dictionary in the following words: "The unlimited or infinitely great three-dimensional expanse in which all material objects are located and all events occur." The Webster's English Dictionary (college edition) puts it this way: "(a) A boundless three-dimensional extent in which objects and events occur and have relative position and direction, (b) Physical space independent of what occupies it, called also Absolute space." That is the general idea, hence no more quotes.

But such definitions give rise to a fundamental question. Expanse of what? When we talk of physical space, everything physical has dimensions. There is no doubt about that. But Einstein is reported to have said that space has no physical attribute. Can there be anything physical without any physical attribute? Generally speaking, people, even scientists, talk of space in a manner as if it is a container. But any container can hardly be boundless. It has to have bounds, otherwise it is no container. And last but not the least is the aspect of space being three-dimensional." Is it really three-dimensional scientifically?

The expanse under consideration is the expanse of energy pervading in all directions. The energy, by its very nature, is neither physical nor visible. It exists, and its existence is the most certain scientific fact.

Any mention of physical energy in modern science denotes a transformed state. The noun space simply denotes distance between any two or more points and beyond. The points are all located within the expanse of energy. If there are no points there would be no distance. In that case there would be only energy. What we see and experience as distance or space is a field created by energy in motion, the field of relative existence. And this field is four-dimensional, not three, time being its fourth dimension, as per Albert Einstein.

This discussion leads up to the conclusion that what we consider as space is a four-dimensional field of energy in motion or the kinetic energy in which all objects and events occur and have relative positions and directions. And this is exactly what Vedic *Ākāśa* is.

Let us see what the noun *Ākāśa* has to say. It is derived from the root *kāś* to shine brightly. The prefix *ā* conveys the sense of *samanta maryādā*, limit all around. Hence *Ākāśa* is the field of the limit of light all around and as light is always in motion, the occurrence of motion is also included in the said limit. In short in the words of Modern Physics, it is the field of the kinetic energy. According to Sāṅkhya, its attribute is *śabda*. The root from which the attribute conveying word is derived is also *sabd*, to denote *āviṣkāra*, i.e., making visible, manifestation. In other words, *Ākāśa* is the field created by kinetic energy to be the site for the process of manifestation of objects and occurrence of events therein.

As an additional comment it may further be mentioned that the etymological explanation of the term *Ākāśa* is also capable of covering the phenomena of what the physicists call "expanding universe." With the continuous expansion of universe the field of the limit of light emitting from numerous galaxies, would automatically go on expanding farther and the universe would continue to remain, so to say, within the bounds of light. The state of expanding universe has already been covered in the explanation of *Hiraṇyagarbha* in chapter 2.

The K.E. field theory also clarifies another phenomenon related to the Vedic revelation in this regard. Veda states:

tadejati tannaijati taddūre tadvantike
tadantarasya sarvasya tadu sarvasyāsya bāhyataḥ

(Y.V. 40/5)

This *mantra* says that it moves others, it does not move itself. It is very far, it is by one's side, it is the innermost of all, it is also the outwards of every body.

Modern physicists' findings endorse the above Vedic statement. They (the physicists) specifically state that whatever is there in space, is under motion, Nothing is stationary. But the space itself does not move. This is the stated position of Veda as conveyed by the *mantra* above.

The position of Veda is simply correct. It has been explained in chapter 1 that the inertial *Prakṛti* (the energy) remains at rest under the influence of its first attribute, namely, *sattva* or existence. The second attribute *rajas* excites merely a part of the whole, causing that much to take to motion and become kinetic, thereby creating a field for the universe to come. Such a motion is therefore limited to that part which subsequently becomes the substrata of the universe. The motion in a part is just like an oceanic current. The ocean stays as it is. Everything in this field is on the move except *Ākāśa*, if we may be excused for indulging in such an expression, because the bearings of *Prakṛti*, the energy at large, remain unaffected.

That the field of operation creation is comparatively much smaller than that of the state at rest in which *Prakṛti*, the energy, exists, is well affirmed by Veda itself. The creation has been described in Veda in an excellent poetic composition. The relevant Vedic *mantra* states:

pādo'sya viśvā bhūtāni tripādasyāmṛtam divi (R.V. X.90.3)

It says: All manifestations are within its one step, the immortal remains at rest in three steps. Even though Modern Physics has so far hardly uttered a word with regard to this aspect of existence, Vedas are unambiguously and emphatically assertive and do point in a direction which remains still out of reach for the scientists of this age. A *pāda* (step) also means one quarter.

In the first chapter while discussing *rajas*, the second attribute of *Prakṛti*, we have hinted a return to the same. The discussion on space in the light of the Vedic view is the right occasion for the same. We may once again take into consideration the derivational interpretation.

The term *rajas* is derived from the root *rañj*. This root is used to convey more than one meaning. Its main usage is to denote colour. It also means to go as explained in the previous reference. Here we present this analysis on the basis of colour. Both of these meanings are stated in the dictionary.

The derivation of *rajas* from Sanskrit root mentioned above indicates that the area of the field created by motion of *Prakṛti*, the energy, is imbued with colour or colours. That the light waves display different colours at different frequencies is a matter of elementary knowledge in Modern Physics. The shades of these colours are very many, but all of them get divided broadly in seven groups, each displaying one colour predominantly. The people at large witness such an arrangement in a rainbow, while the students of science may see them at will through a spectrum or crystal. Veda mentions seven *aśvas* of the sun. Unfortunately the word *aśva* which conventionally means a horse, is mis-interpreted by the substitution of such a conventional meaning. Derived from the root *aś* meaning to pervade *aśva* denotes light or the wave of light which pervades.

One may question such an interpretation. If *Ākāśa* is the field of light, why is it that at night it appears dark with shining stars strewn across it? Why does it not appear shining, say like a huge bulb or something? The reason is simple. The human eye is not sensitive enough to see a light wave. We see only reflected light. On a moonlit night, when the sun is behind the earth, we do not see any beams of sunlight crossing the space in between and falling upon the moon. Yet the moon has no light of its own. It makes earth-nights pleasant by reflecting the sunlight received by it. The moon is our nearest neighbour. In the immense vastness of *Ākāśa*, there may virtually be innumerable dark bodies like our earth and moon. The reflected light received from them belonging to the brilliantly shining stars, which are suns in their respective fields, is too weak for the sensitivity of our retina. That is why the space, in spite of being lighted, appears dark to us. But the seers of Veda knew that there is no darkness, the entire space is full of light. Hence it came to be named *Ākāśa*, alight to the limit.

Two other nouns which describe *Ākāśa* very characteristically are *Śiva* and *Mahādeva.* These two have become greatly embroiled with the personified deities that came into being in the post-Vedic era in India. Even though personification or presenting an abstract or complex idea or subject by giving it some form is a creative activity of human mind, and that way so many gods and goddesses originating in and forming the body of the so called Indian mythology are artistically a superb presentation of genius, yet the attempts to make the masses believe it all as literal truth and objects of faith and ritualistic worship mislead the entire nation into blind lanes of ignorance and make-belief. But this is no place to go into all that. As a sample case, we just offer explanation of the two terms mentioned above before proceeding further.

The noun *Śiva* may be derived from more than one root and would denote meanings likewise. We take one and leave the rest. It is root śī to lie, rest, repose, etc. With the addition of the suffix va, the derivation *Śiva* means *śete asau śivaḥ*, i.e., the one that is at rest stretching in all directions is *Śiva.* Similarly the name Mahādeva means the great illuminated one. It is *Ākāśa* which is at rest in all directions. Whatever there is within, it is moving, but not the space itself. And the great illuminated one has already been explained earlier. There is none greater than *Ākāśa* all full of light in the universe.

As far as *Ākāśa* having its limits is concerned another reference from the treatise *Bṛhadāraṇyak*opaniṣad may also be worthy of consideration to establish that an interpretation such as the one being presented is the correct mode of interpreting the Vedic terms, particularly *Ākāśa* presently under consideration. The book named here is one of the leading *upaniṣads.* Our reference is in respect of the part of a *Śāstrārtha,* a debate on the interpretation of Śāstra, the book of knowledge, between Maharṣi Yājñavalkya on one side and a whole group of leading scholars of the age on the other, as reported in the said Upaniṣad. The pertinent part of the exercise are the two questions put to Maharṣi by the most eminent woman scholar of that time, Gārgī. She declared beforehand that the two questions would reveal the very depth of her rival's knowledge.

Gārgī asked: O Yājñavalkya, beyond the world of stars above, and also below our planet Earth, and in the space in between the two, that which exists and is to come into existence in future, is called *Jagat* (universe in motion). What is it in which the same is involved?

Yājñavalkya replied: O Gārgi, beyond the world of stars above, and below our planet Earth, and in the space between the two, whatever is said to have come to exist and is yet to come in future is involved in *Ākāśa*.

Repeating her question and the answer given she again asked: In what is *Ākāśa* involved?

Maharṣi replied: O Gārgi, that which is *akṣara,* undecayable, *avināśī* indestructible Brahma, that which is capable of attaining growth through transformation, in that the above said *Ākāśa* is involved.

In short the sum and substance of the questions and answers above is that as great an authority on Veda as Maharṣi Yājñavalkya has confirmed our interpretation that *Ākāśa* is a *maryādita kṣetra* (a field having limits all around).

The capital question that remains to be considered is: What would happen on the doom's day? No physicist with whom we had the privilege of discussing this aspect was even inclined to entertain the idea of space or even the sense of space being no more at that time. Courteously, but without doubt, they expressed the view that some sort of space would exist even when all of the material world is gone.

So far as *Ākāśa* is concerned, in a separate chapter on Pralaya, dissolution, this position of its dissolution would be discussed in greater details. Let it suffice here to say that in the great dissolution, which may be taken as the final one, even Ākāśa would disappear. But this much may be pointed out that space is a relative term and if the relativity comes to an end, space would be no exception.

The Vaiśeṣika treatise declares *Ākāśa* to be an eternal causative element together with *Vāyu*, the force and *Kāla*, the time, but says that during the *Mahāpralaya*, great dissolution, they remain dormant and are not in a state of manifestation, i.e., they disappear.

The *Ākāśa* is said to possess only one guṇa (attribute). Prof. Einstein has said that space has no physical attributes. According to him the only attribute that goes with space is said to be that the electromagnetic waves can travel through it unhindered. The Vedic stand does not recognise it as an attribute. The entire expanse of the so-called space is full of energy, and as electro-magnetic waves are nothing but energy in motion, the nature of the space is in no position to offer any resistance to them. Such a movement is naturally a normal phenomenon.

Veda declares *śabda* to be the sole attribute of *Ākāśa*. It is derived from the root *śabd* to mean *āviṣkāra* and *bhāṣaṇa*. The second meaning is obviously speech in which sense it is widely used. But it is the first, i.e., *āviṣkāra* which is most pertinent in respect of *Ākāśa*. The dictionary shows it to mean making visible, manifestation, which is the sole purpose of the creation of that field. *Ākāśa*, as we have seen earlier, is the site of the entire creative activity culminating in the manifestation of universe as well as the maintenance thereof.

The meaning of the word *āviḥ/āviṣ/āvir*, all the three being one and the same, is extremely pertinent from the scientific point of view. The remaining *kāra*, it may as well be *bhāva*, is simply added to make the compound term an abstract noun, for kāra means been or being done and bhāva denotes being. The term *āvis* means making visible something which is/was already in existence but invisible. As stated earlier in chapter 1, *Prakṛti* itself does not possess any mass, and hence remains in an invisible state. It cannot be seen with the help of even an instrument howsoever fine. But once it starts transforming itself into mass/matter, it becomes visible or perceptible in this field of light called *Ākāśa*.

By the mere use of word *āviṣkāra* to denote the attribute of Ākāśa, Vedas are communicating to the human beings in general and the students of science in particular that the invisible *Prakṛti*, at rest, by becoming mobile as *ṛta*, transforms a part of its expanse wherein all manifestations take place.

So far as the second meaning denoted by the noun *śabda* is concerned, the speech, riding on the electro-magnetic waves becomes capable of travelling to the very limits of the field of light soundlessly. *Ākāśa* is full of them. We don't hear because the human ear can hear only the sound carried to it by waves travelling through the medium of matter. It is noteworthy that *śabda* means *bhāṣaṇa* (speech) and not *śravaṇa* (hearing).

— 12 —

VĀYU

The second *mahābhūta* is *Vāyu*. It manifested soon after the first one, *Ākāśa*, came into being and the process of condensing K.E. into mass/ matter commenced. It should, therefore, be obvious that such a name could not be given to it if Veda wanted it to denote air or wind. *Vāyu* is a causative element, and as per Veda, atoms were still an object of future. So, let us dispel this confusion or misunderstanding from our minds that *Vāyu* is air or wind in this context. Like *Ākāśa*, *Vāyu* too has no material existence whatsoever, although it influences matter all over the cosmos. It possesses two attributes, *śabda* of *Ākāśa* and *sparśa* of its own. These two would also be discussed later in this chapter.

Then, what does *Vāyu* stand for in Vedic Physics? It has well been described as *balinām̐ baliṣṭhaḥ* meaning the superbly forceful among forceful ones. In other words what is being conveyed through such an expression is that there is nothing or nobody more full of force than *Vāyu*. Such a declaration can be made in favour of force only. There can hardly be anything more forceful than the force itself. The force by itself reigns supreme in its own field. Hence *Vāyu* is the Vedic name given to what Modern Physics knows as force. It appeared in the *Ākāśa* at the initial stage of creation and immediately got busy. That is why it is said to have been born of *Ākāśa*.

Modern science concedes the importance of force in creation, and has adopted this very noun to denote it. It also affirms the position that force became manifest at the earliest stage of creation, and that the entire creative activity and the order in the system called universe is under its influence. Newton has described it as an external agent that casts its spell upon all kinds of bodies, big or small, existing in the vast expanse of space. In short the entire cosmic operation is affected by its operation.

Wherefrom did force come to manifest? In this respect there is an interesting fable in *Kenopaniṣad.* The same is being narrated here briefly. A Yakṣa, a spiritual apparition, appeared before *Vāyu* and *Agni* (third *mahābhūta*). He placed a dry blade of grass before them and challenged *Agni* that if he had such an immense capacity to burn let him demonstrate it by burning that blade. *Agni* did his utmost, but failed to burn the same. Then the Yakṣa turned to *Vāyu* saying that if he was so full of force, let him just stir the blade. *Vāyu* too applied all the force he could but failed to have any effect upon it. According to the fable, after this happening, Indra went towards the Yakṣa who by that time had disappeared. The derivation of *Indra* being from the root *ind* to mean supremacy or supreme sovereignty, it is a personification of energy. Hence the fable relates the disappearance of *ṛta* (K.E.) that was the Yakṣa. In this manner making both *Vāyu* and *Agni* realise that the source of their respective force and burning power lay elsewhere, the Yakṣa became invisibile. The lesson is that the real source of power behind the force of *Vāyu* and the burning power of *Agni* is *ṛta*. It has already been pointed out in chapter 4 earlier that according to *Nirukta* the term *ṛta* also denotes electricity. *Indra* as well has such a denotation. They are two faces of energy. That is why, as Indra approached, Yakṣa disappeared.

The same thought is expressed in a *mantra* of *Kaṭhopaniṣad*. The relevant part is *tameva bhāntamanubhāti sarvam, tasya bhāsā sarvamidaṁ vibhāti.* Because of its splendour the whole lot gets illuminated. Its luminosity illuminates the entire universe.

The fable is a typical Upaniṣadic style of communicating the scientific knowledge. The *Kenopabiṣad* is stating that the capability of *Agni* to burn or that of *Vāyu* to stir was due to such capacity vested in *ṛta* (K.E.) which had appeared as Yakṣa. The *ṛta* being *Prakṛti* or energy in motion was already affected by rise in temperature initially inherent within energy at rest at sub-absolute zero degree state. R.V. 10/190/1 quoted in the opening of chapter 4 on *Ṛta* specifies this fact. *Agni*, which will be discussed in the chapter succeeding the present one, and *Vāyu*, right now under consideration, are bestowed with their respective ability to interact, out of the enormous capacity and capability of *ṛta*. The electromagnetic force comes into play when energy is in motion, i.e., in the state of *ṛta*. The Yakṣa disappears as Indra, another name for energy, approaches.

In fact the operation creation is basically the condensation of energy in the form of matter and systematic dispersal of the so created bodies of matter under motion in *Ākāśa*. Even modern science confirms the application of the constructive influence of force everywhere in this regard. But so far it has not succeeded in solving this mystery through the evolution of a single formula, even though attempts are on. That is why cognizance of force has been taken through its classification into four types on the basis of interaction and reach of its influence and the respective strength of operation has also been determined. For a comprehensive grasp of the Vedic Vāyu it would be proper to have a picture of force as drawn up by the modern physicists.

Interaction	Comparative Strength	Range
Gravitational	10^{-40}	Infinite
Electromagnetic	10^{-2}	Infinite
Strong	1	Short
Weak	10^{-13}	Extremely short

Modern science appears to have succeeded in resolving the unity of three of the four kinds of forces stated in the chart. The gravitational force is still eluding falling under a common name. It is hoped that this knot may get unknotted soon. But Vedas are quite positive and speak of only a single name *Vāyu* to denote force inclusive of all the four interactions shown above. We shall see below that *Vāyu* denotes all of the said four interactions. Veda has not given any formula, but has presented to us the answer in the choice of the name *Vāyu* which denotes just the four interactions.

Having seen this sketch of force as understood in the world of modern science, let us revert to our main subject *Vāyu* and see what Vedic Physics has to say about it. The derivation of *Vāyu* is from the root *vā* denoting *gati* and *gandhana*. Gati as stated earlier, indicates knowledge, motion and accomplishment. The suffix applied is *u*. The three meanings indicated by *gati* also disclose that the motion or action towards its accomplishment would also imply the proper direction to be taken. In other words; the purpose is covered under the object to be accomplished, the action for the same covered under motion and the right direction taken or to be taken is covered under the sense of knowledge.

The other meaning *gandhana*, is an abstract noun derived from the root *gandh* with the addition of the suffix *ana*. The root *gandh* denotes *ardana. Ardana*, also an abstract noun formed like *gandhana*, is in turn derived from the root *ard* which is mentioned twice in two different sets of derivations. As an abstract noun *ardana* may be derived from both the roots; its use here shall imply the meanings of both of them. The root *ard* placed in one division denotes *gati* and *yācanā*. *Gati* is already covered under the basic root *vā*; hence what remains is *yācanā* which means solicitation. In the other division the root ard is stated to denote hiṁsā, i.e., violence/duress. To make a long dry analysis short, solicitation and duress indicate the approach to determine the meaning of the noun *gandhana*, the key to reveal the mystery of *mahābhūta Vāyu*.

The *Viśva* lexicon states four meanings of *gandhana*. We quote: *gandhanam sūcanotsāhahiṁsaneṣu prakāśane*. It says *gandhana* denotes *sūcana, utsāha, hiṁsana* and *prakāśana*. These are the four fields denoted to express the meanings of the term *gandhana* or the interactions of *Vāyu*. We explore these four below, each separately, but as we have begun our discussion to present *Vāyu* as a synonym of force in the science of physics our exploration will focus upon that aspect specifically.

Sūcana force: Derived from the root *sūc* denoting *grahaṇa* and *samśleṣaṇa*, the abstract noun *sūcana* means to inform, indicate, take, hold and unify, join in support. Through this force each object keeps others informed of its location for the purpose of attracting or being attracted for unification. This process of information is universal, howsoever weak it may be for being effective. Between two particles each keeps the other informed, but being balanced they stay put wherever they are. In the context of force the said information may mean nothing else but an indication to join. Hence the *sūcana* force is the gravitation of modern times. It is noteworthy that this being the weakest of the forces has a universal reach, which means that each one of the small or big bodies keeps on informing its presence to the whole world, and each, trying to unify under its own circumstances, attempts the moment an opportunity arises.

Utsāha force: The noun *utsāha* literally means enthusiasm to rush forth or away. Such a situation is manifest between the opposite charges of electric current, magnetic poles, which rush forth, while in a situation of likes facing each other they rush away or repel. Again, in the event of an electric charge taking to motion the instant appearance of the relevant magnetic force along the same is equally a demonstration of *utsāha*. In fact utsāha is the basic characteristic of electro-magnetic force. This shade of meaning is equally well contained in the noun *gandhana* which characterises *Vāyu*. This typical interaction is virtually a stamp of the electro-magnetic force.

Hiṁsana force: The noun *hiṁsana* means to hurt or to restrain forcibly somebody to act against its inherent nature. Such a situation of restraining influence exists within the nucleus where the superior strong nuclear force keeps the positively charged protons restrained and bound. It has already been clarified that at each stage of an atomic growth a neutron or a proton gets added to the nucleus. A neutron is a neutral particle, but a proton being a positively charged particle is liable to disturb the nucleus by its very presence among its likes. But it is subdued by the hiṁsana superiority of the strong force ruling within. Obviously this Sanskrit term is synonmous to the strong nuclear force.

Prakāśana force: The word *Prakāśana* means manifestation. Nucleus is the basis of the manifestation of all kinds of matter atoms. The nucleus is occupied by protons and neutrons as nucleons. The positive charge of the protons is the only charge within the nucleus. The negatively charged electrons stay out in the orbit. The strong force has a complete sway over the nucleons within the nucleus. It makes even the repulsive interaction of protons to each other ineffective. Each formation within the nucleus attracts an equal number of electrons in the orbit vis-a-vis the number of protons among the nucleons. And yet the weak nuclear force stimulates decay of a part of the nucleus, under the very nose of the strong force, so to say, disturbing a neutron by expelling an electron from its body, and thereby increasing the number of protons within, resulting in an automatic reciprocal adjustment of the number of electrons in the orbit, and the manifestation of a new atom. Such an interaction has been summed up by one word *prakāśana* in Vedic Physics, effecting just the manifestation of a new body.

In short, the subject may be concluded by pointing out that Modern Physics describes force by four different interactions and names accordingly. The physicists are looking for some sort of a formula which may be common to the four interactions. The Vedic noun for force is *Vāyu* and not four as per the interactions. The Sanskrit language can coin such a word which denotes all of the four interactions, neither less nor more.

The attribute of *Vāyu* is *sparśa*, that is, touch. In addition to its own, attribute of *Ākāśa*, i.e., *śabda* also accompanies it. This simply means that with *Vāyu* in operation the *Śabda* attribute of *Ākāśa* remains unaffected. Its continuance has been described by way of the same being incorporated in *Vāyu*. *Sparśa*, as explained earlier denotes touch, hold as well as joining in support. The epithet *Vāyu* denotes that the characteristic of the force is to interact in this manner, i.e., through touch, and bring about the consequent results. It is also clear that by itself *Vāyu* has no material form. Its own existence is absolutely formless.

The total dimensions of the creation are held within the field called *Ākāśa*. This position has been explained in the earlier chapter. According to modern science billions of galaxies each consisting of billions of stars and planets etc. tied along with them, are on the move within this field. All of these are moving non-stop along their respective specific paths, but there is hardly any disorder anywhere. Such an unprecedented, superb system is governing all the bodies of the universe. The credit for all of this goes to *mahābhūta Vāyu*. The manner in which *Vāyu*, i.e., force has grouped and controlled the heavenly bodies and their respective groupings is unparalled in all respects. The more we get acquainted with this creation the discipline of *Vāyu* becomes more and more evident at each step. We shall revert to this aspect later in chapters on *Kāla* and *Pralaya*.

— 13 —

AGNI

Vedic Physics declares *Agni* to be the third *mahābhūta* which is instrumental in the creation of matter and material universe. The conventional meaning of *Agni* is physical fire that appears on account of burning something. Such a fire needs some material, say wood or coal to burn. But the state of a *mahābhūta* itself is a prerequisite for the creation of matter. The *mahābhūta* precedes it. Hence the physical fire cannot be a *mahābhūta* as the same needs relevant burning material beforehand, and makes its appearance only out of it through burning the same.

Veda names three kinds of *Agni*. The first is the one in the sun and the stars. This, as we know, is the thermonuclear *Agni*. The second is the one we see in an electric spark. It is called electrical one. And the third is the physical one born out of burning some object. The first is called *dyūsthānīya* (sun or star based); second is, *madhyasthānīya* (mid-based) and the third is *pṛthvīsthānīya* (earth-based). There is no other fourth kind of *Agni*. But the term *Agni* is used here to denote a *mahābhūta* and as such is a causative. It must not be confused with the noun *Agni* used anywhere else.

In the creation of matter *Agni* takes over during the period *ṛta* undergoes the process of transformation to form, after *Vāyu* has done its job. That is why it is said to have the two attributes of *Vāyu*, namely, *śabda* and *sparśa* as well as one of its own called *rūpa* meaning image or visible appearance. *Agni mahābhūta* is also known as *Tejas*, sharpness or brightness. It manifests itself in two ways. On the one hand it appears as heat, on the other it takes the form of light. This two-way appearance is having its origin in the thermo-nuclear source, as it imparts heat and light.

Modern Physics presents heat and light as two different subjects. According to a physicist heat is transfer of energy. If a bucket full of hot water is poured into a vessel having water at ordinary temperature the heat of water in the bucket spreads in the entire water, thereby reducing the temperature of water which came from the bucket and raising that already there. But the total heat of the water would remain equal to the total of the two heats spread over the total quantity of water. This is so because the excess heat from the hotter water transferred itself to the lower one. There is no loss of heat, but the increased quantity of water has reduced the temperature.

Physicists have also determined the absolute zero temperature. They observed that within a rigid volume the pressure of a gas displays linear relationship with its temperature. If such temperatures are extrapolated the lines of different gases meet at one point. This is the absolute zero temperature point showing the limit of measurement. It is -273.16 degree Celsius. This does not mean that there is no temperature further below. Physicists concede the existence of temperature but declare the same as unmeasurable. Such a state with respect to heat is considered potential heat. The existence of heat is accepted in science even under absolute zero degree on minus side, for heat is energy itself.

According to Veda *Agni* is eternal. It is ever present. It pervades *Prakṛti* (energy) in a state called by modern science as potential heat but named *paramaiśvarya*, absolute supreme-ness, by Veda as we shall see later in this chapter through a famous *mantra*. It causes by way of *Hiraṇyagarbha* to launch itself as radiation on its way for the creation of universe as already described. Then again during the process of creation it emerges under the pressure of *Vāyu* (force) to carry on atomic fusion process to requisite temperatures to enable its successor *mahābhūta* to fuse various elements. It provides all the dynamism to inertial *Prakṛti* to take to motion and transformation.

Having taken note of the advances made by Modern Physics, let us consider the Vedic term *Agni*. It is derived from the root *aṅg* meaning *gati*, with the addition of suffix *ni* and the removal of *n* from the root. *Gati* indicates knowledge, motion and accomplishment. The suffix also gives the derivation a nominative character.

While considering *ṛta* earlier, it was pointed out that the rise in potential heat within the expanse of Prakṛti, energy at rest, resulted in the manifestation of *ṛta* and *satya*. Ṛta is the energy in motion (K.E.). The entire creation is nothing but the result of the process of transforming *ṛta* into compact form. Such a process of transformation is inherent in its motion. Creation is a consequence of the said motion. Therefore the knowledge of the laws of the said motion is the key to the mystery of creation. The Vedic seers have also maintained that Veda is the book of the said laws of motion and creation. Hence it is most logical and proper that the first of the four Vedas, namely, the ṚgVeda, should open its first *sūkta,* a wise- statement, named *Agni Sūkta* with its very first *mantra* (an instrument of thought) commencing with the noun *Agni*. The said *mantra* is:

agnimīḍe purohitaṁ yajñasya devamṛtvijaṁ hotāraṁ ratnadhātamam

(R.V.-1.1.1)

Before we proceed to explain the contents of the above *mantra*, we feel it necessary to clarify the meaning of Sanskrit term *stuti* which is denoted by the verb *īḍe* in *agnim+īḍe* at the beginning. The root *īḍ* which is in use here has been stated to denote stuti. The conventional meaning that has come to be attached to the word stuti is a praise, extolment, or even an eulogy. There are several verbs which denote *stuti*. Such an interpretation cannot be permissible in explaining any science or scientific situation. Interpreting the numerous *mantras* which carry a verb denoting *stuti* with the application of the conventional meaning as above has proved the vital juncture where traditional Vedic scholars have gone astray. It has prompted interpretation supporting multiple deities in Vedic texts, and that too in spite of Vedas themselves saying at places more than one, that the truth or existence is only one, but the wise men describe it in various ways.

The word *stuti* simply denotes a statement. Another abstract noun from this very root, formed with the addition of suffix *ta* is *stuta* which is the Sanskrit original of the English verb to state. And every statement implies its correctness. In the scientific context and more so in Vedic context the word *stuti* bears the above implication by making it specific in details. It means a statement of the properties or attributes, action/interaction and the exclusive trait in the manifestation of its *ahaṅkāra* (chapter 3). The above sentence is simply an elaboration of the three contents, *guṇa, karma, svabhāva,* i.e., attribute, action/reaction and nature of its manifestation, that are required to be stated.

Let us now return to the explanation of the very first *mantra* from Veda. It says: I state the *guṇa, karma, svabhāva* of *Agni* which is the *purohita*, that leads each cycle of creation or which is always ahead in bringing welfare to all *devas*, is brilliance personified, the *ṛitvija*, a performer of *yajña*, the *hotā*, that which is the taker, eater and distributor of the oblations, and the *ratnadhātama*, one which holds the jewels or values of all kinds to the maximum.

Near about 1927 of the Christian Era, a Belgian astrophysicist presented his theory which has come to be known in Modern Physics as the Big Bang. Later on a Russian born physicist also supported and advanced the said theory. As per the Big Bang theory, the universe began initially in the form of a highly dense cosmic egg of primordial matter. This exploded in a big bang. The explosion scattered its parts all around in the space and caused them to fly away. The parts in course of time consolidated in galaxies and stars. The universe continues to fly and expand under that impact.

We would like to state that the idea of a cosmic egg might have been presented to Europe and the West in 1927, but in India this has been a subject of discussion ever since the Vedic age, and has percolated so much among the masses that a completely illiterate village folk, merely adapt to speaking his dilect, is acquainted with the term even today. The name so very common in this country is *Brahmāṇḍa* (*Brahma+aṇḍa*) literally the egg of Brahmā. In India *Brahmāṇḍa* has always been considered a synonym of universe. But Vedas go in far greater details. The Vedic text also speaks about *Hiraṇyagarbha* already discussed earlier which appears once in R.V. and at three different places in Y.V.

R.V. *Hiraṇyagarbha* has already been discussed. The relevant mantra is:

hiraṇyagarbhaḥ samavartatāgre bhūtasya jātaḥ patireka āsīt sa dādhāra pṛthivīṁ dyāmutemāṁ kasmai devāya haviṣā vidhema.

(R.V. 10/121/1, Y.V. 13/4, 23/1, 25/10)

It says: In the beginning only the *Hiraṇyagarbha* existed. It was the sole master of all that came into being. It held earth and all of the shining heavens above. To that illuminated one we offer our oblations. In chapter 2 the whole of R.V. 10/121 has been explained, which begins with the said *mantra*. Here the *mantra* alone is being presented as it appears in respect of different subjects but all elaborating the role of *Agni*.

In the first reference, i.e., Y.V. 13/4 the *devatā* is *prajāpati* and the *ṛṣi*, seer, is Agni. It simply denotes that knowledge about the *prajāpati* being mentioned in the relevant *mantra* has been gained or acquired through *Agni*. The term *Prajāpati* has been discussed in chapter 6 earlier, but there it relates to the ten *Prajāpatis* born at the sub-atomic stage. Etymological interpretation still being the same, the only difference is that here it does not indicate any of the ten discussed there. It simply denotes a procreator, i.e., atom again of chapter 6. The whole universe is its creation. As the first in the periodic table, the Hydrogen atom is the procreator and the rest are its own creations. So it existed in the beginning of the creation, it bears *agni* or *hiraṇya* within its interior, and it held the material out of which this earth and the world of stars above came to be manifested. The relevant *mantra* also relates that knowledge regarding the existence of the Hydrogen atom was gained through *Agni* itself. In short the state of the Hydrogen atom is the state of a Hiraṇya-garbha as related in Y.V. 13/4.

In Y.V. 23/1 an identical *mantra* introduces us to another *Hiraṇya-garbha*. Here the *ṛṣi* is *prajāpati* and the *devatā* is *parameśvara*. The conventional meaning of the term *parameśvara* denotes the Almighty God, but such an interpretation we are brushing aside as unwelcome, as we are examining the scientific content related to the physical creation. The term *parameśvara* is a compound of two words, namely, *parama*, and *Īśvara*. In Sanskrit *parama* denotes absolute and Īśvara the master or the lord. The compound, therefore, means the absolute master/lord. Our prime subject being *Agni*, we have to find a stage where *Agni* is the absolute master.

How about the potential heat below absolute zero temperature? That is the *devatā* named *parameśvara* in Y.V. 23/1. It is *Hiraṇyagarbha* because the *agni* exists in its interior as energy. And it undoubtedly held within itself our earth as well as all the galactic material above. It is noteworthy that the *mahimā*, i.e., the greatness of agni existing in the said state of sub-absolute zero temperature has been specified in the *mantras* related to the same *devatā* that follow. We may add that the knowledge of this *Hiraṇyagarbha* has been revealed or gained in the relevant *mantra* through the seer *prajāpati*, i.e., the Hydrogen atom. It is noteworthy that this aspect has already been clarified in our interpretation of R.V. 10/190/1 in chapter 5 wherein it has been explained that *ṛta* and *satya* got manifested due to increase in heat and temperature within *Prakṛti*, energy at rest. The word denoting heat/ temperature in the said *mantra* is *tapas*.

Now, last but not the least, is the third reference Y.V. 25/10, where the *devatā* too is named *Hiraṇyagarbha*. This having been discussed at length earlier is the primordial ball of fire. The stellar world that followed carried replica of *Hiraṇyagarbha* everywhere.

In R.V. 2/7/6 *Agni* has been called *sahasasputra*. The *Vedic Koṣa* explains the same as: *baliṣṭhasya vāyoḥ putra iva vartamānaḥ*, i.e., existing like the offspring of the strongest Vāyu. *Agni* originates and exists in this state within *Hiraṇyagarbha*. The force of gravity is *Vāyu* under the influence and pressure of which within the body matter *Agni* performs. That is why it is likened as an offspring of the former. The *Nirukta* also describes *Agni* as the offspring of *bala*, force, i.e., *Vāyu*.

As the middle *mahābhūta* the role of *Agni* is to keep on generating increasing heat thereby raising temperature within the confines of the stellar furnaces, under the ever increasing pressure of gravitation, *Vāyu*, to destabilise the particles or their combinations thereby preparing ground for further atomic fusion, which is accomplished by the successor *mahābhūta* in a highly measured way. That way *Agni* plays a dominant contributory role as a causative in the process of creation.

Photon is the particle of light. According to Modern Physics the electron orbiting the nucleus, when it gets excited, moves into a higher sub-orbital. It has been pointed out earlier that there are four sub-orbitals, called s,p,d,f, by modern science, and the so called *mukha* of *Brahmā* to the Vedics. Depending upon the excitement the electron may jump up to any position above where it recovers its balance by releasing excess energy in the form of a photon and jumps back to a lower sub-orbital. For a clearer understanding the sub-orbitals may be compared with rungs of a ladder or steps of a stairway. Like the upward jump, coming downward too may not necessarily be rungwise or step-wise. It may skip a rung/step or two. As it comes down the excitement being caused by the heat catches and enthuses it again and the electron continues jumping up and down at an incredible speed, each time releasing photons. Thus photons are packets of heat energy released as particles of light by electrons.

Modern science as well as Veda tell us that photons move in a ray, which is like a dotted line, the interval between two photon dots being too small and frequency too fast, for our retina to register. That is why a ray appears to move in a line in space, or a bulb appears to keep the field of its range constantly illuminated. Modern Science has developed and perfected this process on an unprecedented scale all over the globe to provide electric light and power, but so far as light is concerned the basis is the same as stated above. It may be reiterated that photon is the only elementary particle which has no mass or matter and is a quantum of electro-magnetic radiation.

Out of a very wide range of electro-magnetic waves varying in their length and frequencies, a very small section falls within the range of the human eyes, and that is why they are called light waves. All electro-magnetic waves, being energy waves are carriers of light but are invisible as our eyes are not sensitive enough for them. Even the visible ones are not directly visible. They are visible only when they illuminate some object or body and get reflected to our retina forming a reflection upon the latter which our brain registers as a view.

The Vedic seers were fully wise to this phenomenon of vision. They knew fully well that light waves or rays are not visible to human eyes. They knew that what we see is the reflected image of the field or object in our vision. Hence they were aware of the scientific phenomenon that whatever is viewed is not light but the image created upon the retina by the reflected light. What we see is not the object or the field, but its appearance or reflection as created upon our retina. This image reflection, *rūpa*, is the attribute of *Agni*. In Sanskrit language the root *rūpa* is given to mean *rūpakriyā*, i.e., making or creating image. What does that communicate?

Veda states seven colours of light. Modern Physics has learnt immensely through the spectrum showing the seven colour groupings. With regard to the subject of vision that we are discussing, the object or body illuminated by light ought to have been seen drenched in seven colours, but it does not happen. The viewer feels that the colours that are appearing in his eyes of the object under the light are its true colours. But the reality is quite different. Only the colours that are reflected are being received in the image formed in the eyes. A rose flower appears red and its leaves green to the viewer because the remaining colours of light have been absorbed by the flower and the leaves respectively, and only the colours being reflected are colouring their images within the eyes. This is the state of this phenomenon in respect of all the images that are being received by the viewer. An African is black because his skin absorbs all the colours of the light. Whatever our eyes see is not the true image of the reality. The image does not reflect a true picture. It is the appearance as created upon our eyes by the reflected light waves that forms the image. The creation of such an image is, therefore, the attribute of *Agni*. The first half of the last *mantra* of *Yajurveda* reads:

hiraṇmayena pātreṇa satyasyāpihitaṁ mukham

The *mantra* says: With the recepticle of light the face of matter is hidden. In a few words the *mantra* has said all that has been described in the preceding para; in short, what has been stated as the creation of the image. The reality of the matter remains concealed. It has already been clarified in chapter 5 that Vedic noun for matter is *satya*. Prof. Einstein has established his relativity on the basis of motion. We shall discuss that later. But Veda here is telling us that even the vision is relative. That is why the entire thrust of Veda is to stress upon the *homo sapiens* that the appearance must be taken *prima facie* only. The reality must be determined through analysis and testing, etc. To know the appearance may be acquaintance, but to find out reality is science.

In the creation of image, apart from the light aspect of *Agni*, its heat side also contributes simultaneously. The light as well as the heat are both energy radiation. How the light creates image has been mentioned above. The heat affects this phenomenon in a different way. It causes changes to occur in the present state of matter. The greenery of the vegetation appears to turn yellowish. The heat causes reduction in the fluids within; they start drying up. The dried up leaves start scattering away. All of such changes affect the appearances of the reflected images. The heat has a burning effect which in turn brings about changes in the appearance of its images. The changes in the state of vegetation are simply one such illustration of this comprehensive interaction. In fact the heat effect continues to affect the creation of image phenomena accordingly all over. The natural phenomenon is that the creation of image interaction results from the reflection of light from the heat affected objects or matter. Hence Veda states *rūpa* image/appearance as the attribute of *Agni*. The other two attributes, namely, *śabda* and *sparśa* of *Ākāśa* and *Vāyu* respectively continue during the operation of the third *mahābhūta* as well. That is why the two are also mentioned along with the third one.

The third *mahābhūta* Agni performs a positive role in the process of creation, but in the process of dissolution its contribution is supermost according to Veda. At this stage of our presentation of Vedic Physics we are discussing the contribution of the *mahābhūta* to the stage-wise creation and development of the matter, hence this has been confined within that limit. The list of creative elements enumerated by *Sāṅkhya* (at 1/61) does not mention dissolution, as the latter is not a creative element. It is the process of undoing the creation. But Veda presents both creation and dissolution as two steps of the continuous solitary motion. That is why an additional chapter is being added on dissolution after the completion of Sāṅkhya-narrated elements. The role of *Agni*, being stated therein at length, has been left out here. Our own view is that without grasping the dissolution part of Vedic Physics its present part related to the creation remains incomplete. The dissolution process is no less a scientific phenomenon than that related to the creation. Probably such a view is the inspiration behind a number of modern physicists and scientists who are paying their attention to it and presenting their views in that respect.

— 14 —

JALA

Jala is the name of the fourth *mahābhūta* which follows *Agni*. The conventional meaning of this term is water, but in this context it is hardly applicable. The misapplication of the said conventional meaning in the past led even the scholars astray and away from the Vedic knowledge. This has been pointed out earlier as well, but it is also a sure indication of the fact that the knowledge born out of the Vedic studies was prevalent in India wherefrom it spread westwards at least as far as Europe, maybe under a mistaken interpretation. As has been pointed out earlier, a *mahābhūta* is instrumental for the creation of matter. As such it is required to register its presence prior to the appearance of matter, and water itself being a material product, cannot be taken as one. The role of *Agni* is to destabilise a stable combination. The process of manifestation could well proceed through a successor that may re-fuse the destabilised particles in new combinations. And that is the interaction *Jala* brings about.

According to modern science the building blocks of the atomic structure throughout the universe are three elementary particles, namely, electron, proton and neutron. The two opposite charges, i.e., negative and positive, imbue the three particles, electron and proton each exclusively but respectively, and both of them together counter-balancing each other thereby neutralising the neutron. The atomic growth by way of fusion gets initiated, the first step being the deuterium in which a neutron steps into the nucleus to join the proton. Henceforth irrespective of the process, the manifestation of all the elements of the Periodic table, including those not available in nature, but developed in the scientific laboratories, as well as their isotopes, follow through the process of the addition of a neutron or a proton in the nucleus, the latter with an electron in orbit. Moreover the number of neutrons within any nucleus cannot be less than that of the protons. The former must equal or outnumber the latter.

It is noteworthy that the mass of the first atom, i.e., the Hydrogen one, is equal to the mass of a proton plus an electron. The mass of a neutron is just about the same. Moreover a free neutron within a short time decays into an electron and a proton. When a neutron enters the nucleus, it brings about merely a physical change in the atom the nucleus of which it has entered, causing the nucleus to become heavier. But when the two particles join in separately, i.e., the proton going into the nucleus and the electron out into its orbit, the new formation results in producing both a physical as well as a chemical change, the latter amounting to the manifestation of a new element of matter, an atom.

It is clear from the above narration that the development of atoms from hydrogen state proceeds in definite steps, a definite amount of mass equal to the total mass of a proton and an electron gets added to the mass of the developing atom at initial stage, either in the form of the duo, or by way of a neutron. Only that much gets added, neither more nor less. It is also not difficult to understand that the chemistry of an atom rests with the number of electrons orbiting the nucleus. That is why each time the additional mass gets unified as two separate particles, the proton going within the nucleus, and the electron into the orbit, equalising the number of electrons with protons, the chemistry also changes and a new atom gets manifested. But when the electron enters the nucleus as a part of the body of a neutron, without affecting the orbit, the increase of mass merely reflects in its physical state, and the chemistry of the atom remains the same. That is why it is called an isotope.

Such an addition of the mass is kept strictly within the limitations of either of the two options, i.e., the neutron state or the separate state of a proton and an electron. If re-actants coming together add up to an increased amount of mass, the designated mass is retained and the additional mass, if any, is discarded. For example, during the manifestation of the Helium atom, the extra mass gets thrown out. Naturally such a mathematical determination of the mass just equal to that of a Hydrogen atom, retaining the same within the nucleus in case of an isotope, or another proton in the nucleus and the lighter negatively charged electron in the orbit for a new atom and discarding whatever mass is in excess, can hardly result from the interaction of *Agni*. It shows that another instrumental factor is at work which has been well taken cognizance of and named *Jala* by Veda.

The process of decay is observed elsewhere as well resulting in manifestation or transformation of matter. Modern physicists are well acquainted with the alpha, beta and gama decay. The alpha decay is in fact decay by way of a Helium nucleus matter at a time and the beta decay is the decay by an electron matter. Similar is the state of the decay by way of a gama ray which occurs by way of electromagnetic energy release in consonance of a photon. The radioactive matter bears its name because it decays. Hence decay is also a process of the manifestation of matter.

In the light of the above narration based on the findings of modern science, let us see what the noun *Jala* has to say. The root *jala* in Sanskrit appears twice, each differently. We take up the one that has a bearing on our quest. It is stated to be used for *ghātane*, i.e., to kill or destroy. As in physics matter or energy neither gets killed nor destroyed; it is clear that in this context it has to be interpreted by way of transformation which means end or death of one form and the emergence of the other involving the matter of the former. Hence *Jala* causes the manifestation of a new form out of the matter and energy available at each step of the creation. Thus giving form to the particles, isotopes or atoms highly excited under the influence of the third *mahābhūta Agni* in accordance with the pattern set by the process is what the *mahābhūta Jala* has to perform. The prefix *sam* added to *ghātane* denotes fusion, while *vi* communicates fission or break up. Thus the term *ghātane* covers both aspects.

It may be understood that *Jala* has no form or figure of its own. It remains formless and interacts that way only. The specific change that is brought about or effected is the proof itself that *Jala* is at work. In other words, its work is its identification. Its entire performance is carried out at the energy level. It emerges within the *Hiraṇyagarbha* or in the stellar furnaces or even within the scientific labs of the modern era inside atomic re-actors. As a *mahābhūta* it is eternally present in all forms of matter, and governs all fusions and fissions.

It may also be pointed out that Modern Physics is now fully wise to the phenomena that began with the Hydrogen atom commencing to grow through re-adjustment within itself of the mass of matter equal to its own multiple at each stage or step thereby transforming itself into an isotope or a new atom through fusion/fission, but it has not so far been able to explain any basis for the same. Hence the physicists accept the whole process as the result of the cooking that went on and still goes on within the intensely hot interior of the primordial fireball or its parts subsequently consolidated as stars. But Veda identifies this under separate interactions, this one under the name *Jala* which denotes accomplishing new forms under heat effect through fusion or fission.

This narration of the performance of *Jala* in bringing about the manifestation upto the stage of atoms, will remain incomplete without mentioning a by product of this creative process. It has already been pointed out that an atom remains neutral as it has an equal number of positively charged protons in its nucleus counter-balanced by negative electrons in its orbit. But circumstances do get created when the number of the electrons in the orbit may get affected, resulting in a tilt in charge balance. In such a case the atom shall become charged. Such charged atoms are called ions. The manifestation of the matter in the form of atoms or ions, by fusion or fission is the effect of the *mahābhūta Jala*.

Modern Physics has successfully conducted indepth study, based on practical experimentation of this natural process. Modern scientists have discovered that all manifestation is that of energy in the form of matter. They have found out that at each step of manifestation, whether that of an isotope or an atom, the particles composing the body of the reactants transform into the new reaction product as its successor. They found that at such a transformation the particles make new adjustment amongst themselves which result into the emission of energy. They have grouped the atoms into two groups on that basis. In the first group come the atoms mentioned in the upper part of the Periodic Table.

Their manifestation results in the emission of energy at each step. In the second there are those atoms where the reactants have less energy and the end reaction product requires more energy for its manifestation. The first is called the exoergic reaction group of atoms and the last one the endoergic reaction group. The first group is also known as carbon group, with a preliminary stage ending at the Helium atom.

Vedas have specifically indicated this state of affairs in R.V. 1/163/9 already explained at length in Chapter 8. The said *mantra* describes a supernova and mentions therein the accomplishment of fusion upto iron-stage. The iron stage is no emission stage. The *mantra* narrates its extraordinarily brilliant burst which is described by the Veda as *Hiraṇyaśṛiṅgaḥ*, the peak of brilliance, during which the fourth *mahābhūta*, *Jala* avails an unusual stage to finalise transformation of elements known as endoergic.

Returning to the matter of experimentation modern science has already developed the *modus operandi* for producing nuclear energy both through the fusion as well as the fission process. We have now a number of nuclear power states. But apart from its strategic importance, it has tremendous scope for peaceful application. Through numerous experiments the emission of energy through fission has been worked out. It has been established that when a heavier element is struck by a neutron fired at it the heavier element breaks up into lighter nuclei. For example a Uranium 235, so struck by a neutron causes the unstable Uranium to be effected by getting divided immediately into two lighter nuclei of Barium 141 and Krypton 92 and an excess emission of 3 neutrons. The total mass before and after the division remains the same 235+1=141+92+3. But the reaction resulting from the strike causes emission of energy from the atom struck. Technique has been perfected to effect a chain-reaction whereby much more energy becomes obtainable. This is called fission. The fusion reaction has already been described above.

Attention is again drawn to R.V. 1/163/12 wherein Veda speaks of this possibility in following terms: **"ajaḥ puro nīyate nābhirasyānupaścāt".** It says: "The birthless leads to its nucleus from the fore and later from the back". The birthless is the energy. It leads to its nucleus from the fore, i.e., the process of fusion, and later on from the back, that is, approaching in reverse by breaking the elements which effect energy emission.

It has already been pointed out that modern science has made vast and in-depth study of the natural process of creation and development of the atomic matter. Science realised that emission of energy occurs in both the cases of fussion and fission and having learnt the process as occurring in nature, further succeeded in producing atoms beyond those available in nature. Science created conditions in the laboratories and successfully brought about the manifestation of atoms beyond No. 92, the last available in nature and shown on the Periodic Table. All of these artificially created atoms are radio-active. The number of isotopes coming up in between is far greater. The nucleus of No. 92 Uranium atom contains 92 protons and 143 neutrons forming a ball-like shape. This number increases very considerably in the lab created atoms, their nucleus getting heavier and heavier at each step.

The three attributes related to the preceding *mahābhūta* continue to be effective. As such they have been associated with *Jala* as well. But its own attribute is stated to be *rasa*. It is derived from the root *ras* which denotes *śabda*. It may be recollected that *śabda* is the attribute of the first *mahābhūta Ākāśa* as well. It has been explained there that it denotes *āviṣkāra* or creation. But instead of stating or repeating *śabda* Vedic Physics prefers to use the noun *rasa* which is indicative of some purpose. Let us first clarify the sense of *śabda*, which means creation. With first *mahābhūta Ākāśa* the use of this noun to depict its attribute is to convey the sense of creation. *Ākāśa* as a field is to be the site for creation. But the use of the noun *rasa* is to denote the product of creation. Moreover derived from the second root the noun *rasa* would denote the essence also. In the context we are considering, the word *rasa* would further underline the meaning of the product of creation by adding that it is as well the essence of the specific effort of creation. In other words the transformation has been finalised.

In fact Modern Physics does not take cognizance of any such action or reaction created as stated by Vedic Physics. It speaks of the space which, maybe in a different light, is stated as *Ākāśa* by Veda. It mentions the force which has been named *Vāyu* by Veda, and it details the heat which is named *Agni* by the latter. After that the development of the isotope and the atom is presented in a manner as if it is the performance of the intensity of the heat itself. But that heat by its very nature can be effective in exciting and destabilising the particles or the reactants, but the reaction, particularly a choosy reaction by way of fusion affecting the same reactants is hard to explain through the heat. An unknown factor is at work. The decay in products already created, the discarding of excess mass coming together with the reactants need some explaining and the Vedic explanation is the causative *Jala*. It is an advanced state on the manufacturing assembly line.

The fourth *mahābhūta Jala* is no less active in the labs of modern physicists all over the world wherever they have the plant or reactors working. The chain reaction process is being improved and the nuclear or thermo-nuclear power being generated is the *Alladin's Jinn*. It is also the most effective detergent from the angle of defence. Looking at the energy consumption and the limits of the energy sources available the emphasis is shifting to invent safe processes of employing the nuclear energy in the service of Man by developing radiation safety devices. Everywhere the *mahābhūta Jala* is at your service.

— 15 —

PṚTHIVĪ

The fourth *mahābhūta Jala* completed its operation to the extent of transforming the *tamas* affected *ṛta* (K.E.) into the form of atoms etc. Beginning with the Hydrogen atom the energy consolidated in the physical forms of neutral atoms and charged ions as products created during its developmental growth which continues to come into other material forms, but in both the atomic and ionic stages the state of being is still not quite stable. To bring about the stability in its creation the process takes recourse to effect expansion by tying up its products through effective chemical bonds or combining the elements into compounds. The fifth and the last *mahābhūta Pṛthivī* is instrumental in such a performance. The reaction product after the chemical bond is called a molecule by modern science and a compound is a pure substance made up of elements more than one.

The study of molecules, chemical bonds and compounds etc., falls under the science of chemistry. But as the process of creation extends itself to the generation of molecules and compounds to that extent it comes under the domain of physics as well. As we are discussing the subject of physics as stated by Veda, and *mahābhūta Pṛthivī* being the fifth and the ultimate instrumental operation, its presentation is equally important.

As stated above a molecule is an expanded image of more than one atom and ions bound together through a bond, thereby presenting an appearance of bound matter. The matter resulting from such transformation of energy achieves better stability. In other words, a molecule is more stable compared to solitary atoms. Obviously the nature in this manner assures durability in its operation of creation of universe.

Both molecules and compounds are combinations of atomic elements, the latter on a larger scale. In a molecule two or more than two atoms are combined by chemical bonds. The chemical binding is related to the electrons of the atoms combining. The electrical attraction between the electrons of two atoms occurs in three ways: (1) positively charged atom+negatively charged atom, (2) negatively charged atom+negatively charged atom, and (3) positively charged atom+positively charged atom. The first is called an ionic bond, the remaining two are co-valent bonds. In the first an electron gets transferred from one neutral atom to the other neutral one, thus making both the atoms oppositely charged resulting in a bond between the two. In the other two the electrons are shared by the two atoms. The third one is seen primarily in metals. Such a sharing of electrons may not be equal, thereby providing a symmetrical or asymmetrical charge distribution between the combining atoms.

A compound is a pure substance made up of a combination of more than one element. The composition of a compound may be chemically decomposed into its components. By decomposition the element desired may be obtained by removing others. Many elements are gained from natural compounds, such as metal ores.

According to Veda and Sāṇkhya, after the completion of the operation of the fourth *mahābhūta Jala*, it is the instrumentality of the fifth *mahābhūta Pṛthivī* which combines the products of its forerunner into molecules and compounds to conclude the transformation of energy into matter. For a better and clearer understanding of the Vedic presentation we begin with determining the true meaning of the term *Pṛthivī*.

As seen earlier, the conventional meaning is no help. Conventionally *Pṛthivī* is a Sanskrit synonym of our earth. That cannot be, as there was no earth in the beginning of the creation. The Vedic meaning is obtainable through analysing its derivation as we have seen all along. Let us follow the same method.

The noun *Pṛthivī* is derived from the root *prath* which, though prescribed for a different meaning, carries the note *Vede vistāre*, i.e., in Veda to expand or to enlarge. As we are dealing in its Vedic use the root denotes expansion. To remove any doubt in this respect two authorities are being quoted here. "*(yad) aprathayaṁ tat pṛthivyai pṛthivītvam*". (Tai 1/1/3/607). The meaning is: what was expanded that was the *pṛthivīness* of *pṛthivī*' The other quote is: "*tāmaprathayat sā pṛthivyabhavat*" (*Śatapatha* 6/1/1/15). It means she was expanded, she became *Pṛthivī*. Both the quotes carry the same root coveying expansion being the criterion for the noun *Pṛthivī*. The first quote is naming *prathana*, expansion, as the *pṛthivī-ness*, while the second is communicating *prathana* to be effecting *Pṛthivī* itself. In his famous treatise *Satyārthaprakāśa* Maharṣi Dayānanda Sarasvatī also explains: *yaḥ parthati sarvam vistṛṇāti tasmāt saḥ pṛthivī*: that which expands all, hence it is *Pṛthivī*. Of course, in these quotations it is the noun *Pṛthivī* which has been explained. This noun there has not been reserved to denote the *mahābhūta* under reference. But on that basis the same noun has been established in this context as well. A brief sketch of the combining of molecules and compounds presents before us what expansion of atomic material is. The placement of *mahābhūta Pṛthivī* after *Jala* has been described as completing the creation of matter as atom leaves little room for any other inference.

In fact, we humbly submit that the very noun *Pṛthivī* derived from the root *prath* to expand also has the potential to reveal itself as a seedling of the science of chemistry as its characteristic operation covers the field of the latter too. In any case, this *mahābhūta*, like its forerunners, operates at the level of energy and as such remains itself formless.

Even though *prathana* has been explained, yet it may be of interest to note what Veda says about the molecule of water. Beneath we take up three quotes. Thanks to modern science, the fact that a water molecule is a combination of Hydrogen and Oxygen atoms is almost a commonplace knowledge. Now let us see what Veda says:

mitraṁ huve pūtadakṣaṁ varuṇaṁ ca riśādasaṁ dhiyaṁ ghṛtācīṁ sādhantā

(R.V. 1/2/7)

The *devatā*, subject matter, of this *mantra* is *mitrāvarunau, Mitra* and *Varuṇa*, a dual number. The *mantra* says: **(pūtadakṣaṁ mitraṁ)** Hydrogen which is an expert in cleansing **(riśādasaṁ varuṇaṁ ca)** and Oxygen which eats away by rust formation **(huve)** I accept **(ghṛtācīṁ dhiyaṁ sādhantā)** both of them accomplish the creation of water.

In this *mantra, mitram* Hydrogen has been qualified with the adjective *pūtadakṣam*, that is, expert in purifying. It is a special property of Hydrogen. It purifies metals. If Hydrogen is passed through copper oxide, it combines with O_2 and changes into water steam, leaving behind the residue of pure copper. Hydrogen has been described as the main reducing agent. On the other hand Oxygen makes metals rusty, and rust eats into the metal. That is why *varuṇam* has been described as *riśādasam*, one that eats into the metal with rust. This is the oxidising agent. Hence *varuṇam* is Oxygen which is being qualified by Veda with an adjective meaning that which causes rust. In Sanskrit language both the roots *riś* and *ruṣ* denote harm or damage. The Vedic term *riśādasam* is derived from the first root, i.e., *riś*. The English word rust is a virtual derivation from the second root *ruṣ* with the addition of the suffix ṭa. Only the pronunciation is slightly changed. But it can certainly claim its origin from the Vedic language. The conclusion given in the *mantra* that Hydrogen and Oxygen combine to produce water is being confirmed by modern chemistry.

Not only that. Elsewhere Veda has stated that the combination of Hydrogen and Oxygen results in water with the aid of electricity. Here is the *mantra*:

vidyuto jyotiḥ pari sañjihānaṁ mitrāvaruṇā yadapaśyatāṁ tvā
tatte janmotaikaṁ vasiṣṭhāgastyo yattvā viśa ājabhāra
(R.V. 7/33/10)

The *devatā*, subject matter, of the *mantra* is *vasiṣtha*. Derived from the root vas to reside with the suffix *iśṭha* this noun has been used here to denote water because water is one of the basic requirements for establishing residence. The mantra says: **(vasiṣṭha vidyutaḥ jyotiḥ pari sañjihānam yat tvā)** Hey you, water, a basic requirement for habitation, the one who, under the charge of lightning, changing its form, **(mitrāvaruṇā apaśyatām)** is seen by Hydrogen and Oxygen that its birth is from them **(tat te ekaṁ janma);** that is why one of your name is *janma*; **(uta yat tvā agastyaḥ viśaḥ ājabhāra)** the one which was given to the people by the sun that you bear the name *janma.*

For the purpose of clarification it may be added that in the above translation English word *birth* and Sanskrit noun *janma* are synonymous terms. As we are presenting this matter in English, the word *birth* has been used where such a phenomenon is implied, but Sanskrit term *janma* has been used as such because here the same appears as a proper noun. In the long list of the synonyms of *udaka* (water) given in *Nighaṇṭu*, one is *janma*. The *mantra* apart from whatever else being said, is also stating as to how the synonym *janma* came to be included. The *mantra* emphasises that the basis for such a nomenclature is that it is born out of a union of Hydrogen and Oxygen. Sāyaṇa in his commentary has explained the portion *mitrāvaruṇā apaśyatām* as entertaining the thought that it is born of the two of us. We have taken the same line. It is noteworthy that Agastya has been used for the sun because it destroys many ills. The word *āgas* meaning every ill, *āgas* + *asta* = *agasta* means one that ends all ills. The English term august may also claim origin from Veda as it denotes a sense of grandeur.

It may be interesting to have a look at the very next *mantra* to the last one. The same runs:

utāsi maitrāvaruṇo vasiṣṭhorvaśyā brahman manaso'dhi jātaḥ
drapsaṁ skannaṁ brahmaṇā daivyena viśve devāḥ puṣkare tvādadanta
(R.V. 7/33/11)

The *Devatā* of this *mantra* as well as that of the *mantras* numbering 12, 13 and 14 is the same Vasiṣṭha of the preceding *mantra*. This is a clear indication that the subject matter of all of these *mantras* is water. But we shall limit our presentation to the 11th *mantra* quoted above, and move towards concluding this chapter regarding *Pṛthivī*. The *mantra* says: **(vasiṣṭha uta maitrāvaruṇaḥ asi)** Hey you superior support-resource for habitation, you are the offspring of *Mitra* Hydrogen and *Varuṇa* Oxygen. **(brahman urvaśyāḥ manasaḥ adhijātaḥ)** Hey, grain-donor, you are born of the capacity of electricity/ lightning. **(drapsaṁ skannaṁ tvā)** Transformed into water **(daivyena**

brahmṇā) to cause grain for wise men, i.e., to grow grain for them **(viśve devāḥ)** the sunrays **(puṣkare adadanta)** placed you in the atmosphere.

On a point of explanation it may be stated that the word *brahman* above has been given as a synonym of *anna* (grain) in the *Nighaṇṭu*; *drapsam* is water because it is stored, retained by living beings and is for drinking. Root *dhṛ* to hold, retain + root *psā* to eat, drink = *drapsa*. Some commentators have interpreted it wrongly as the semen, but that cannot be as it is not something to drink. *adadanta* is derived from the root *dada* to give away; *puṣkara* is derived from the root *puṣ* to hold, retain with the addition of the suffix *kara* and is stated to mean *antarikṣam*, the middle space called atmosphere. *urvaśī* is derived from the root *aś* to pervade with the prefix *uru* and the feminine suffix Ī, thereby meaning "that which pervades", a typical characteristic of electricity/lightning. It is specifically pointed out that the prefix *uru* does not mean shanks. The synonym for shanks is *ūru* and not uru. The noun *urvaśī* also goes for a feminine character, but a woman has got nothing to do here in this *mantra*.

The teachers of Modern Physics have been prompting their students to conduct an experiment in the science lab, in which a test-tube filled with Hydrogen and Oxygen is exposed to repeated electric sparks. The reaction changes the atoms of both the gases to a nascent state in which they, becoming highly excited, combine by forming a bond and transform into a water molecule. In the light of such an experiment being conducted by the students of science, we would simply like to point out in respect of the *mantra* under consideration, that all that has been left out from saying is any mention of a test-tube, as the nature's lab has no such equipment. Otherwise the *mantra* is just describing the very same idea in its own way, which harbours such a practical experiment by the students.

Let us close this discussion on *Pṛthivī* with an analysis of its attribute. During operation Pṛthivī, the attribute of *Ākāśa*, i.e., *śabda*, that of *Vāyu*, i.e., *sparśa*, that of *Agni*, i.e., *rūpa* and that of *Jala*, i.e., *rasa* remain active as usual. That is why it is said that apart from its own attribute the attributes of all the four forerunners are also contained in the *mahābhūta Pṛthivī*. The independent attribute of this last *mahābhūta* is *gandha*. What is this *gandha* ? *Gandha ardane*, i.e., root *gandha* is used to denote *ardana*. Again the same question, what is *ardana*? The root *ard* from which the abstract noun *ardana* is derived appears twice. These very two roots and the derivation of *ardana* from both of them has been discussed and presented at length while explaining the derivation of *gandhana* in respect of the second causative, *Vāyu*. It will be a dreary repetition. Those interested in it may turn back a few pages. We proceed here by simply reiterating the conclusion. *Ardana* or finally *gandha* would denote any or all of the meanings stated in respect of both the roots namely, *hiṁsana*, duress, and *yāñcā*, solicitation.

In the operation *Pṛthivī* formation of bonds is the typical interaction between the combining atoms. A bond implies violence or forcible restraint. But it is rather solicitation of such a state which appears supermost, because for any bond to come into being at least the electrons of two different atoms have to be complementary with respect to their protons. Only then the two atoms would form a common bond and a molecule would be formed. Also the state of energy prevailing with respect to the affected electron or electrons would get more stabilized. It is the inherent search for stability that the idea of solicitation appears to be more pre-dominant, even though forceful restraint is always there.

A different derivation of the word *gandha* may also support the operation of this *mahābhūta*. The conventional meaning of *gandha* is smell. But in Veda we seek interpretation through the process of derivation. The word *gandha* has two parts, namely *gam+ dha*. By adding the suffix *ḍa* to the root *gam* meaning gati, knowledge, motion and reach, you get the word *ga* which denotes any object or being in motion. Its objective case would be *gam* which when combined with root *dhṛ* to hold with the suffix *ḍa* would deliver our *gandha* denoting that which holds together two or more atoms in motion, in a combined form, i.e., a molecule. Hence the noun *gandha* expresses most fully and comprehensively the attribute of *Pṛthivī*.

Before closing this chapter it would also be proper to explain the Sanskrit synonym of molecule, which is *Pīlu*. Derived from the root *pīl pratiṣṭambhe* [prati=against (something) ṣtambh=connected with]. In this context (something) is an atom or ion.

— 16 —

PURUṢA - THE EFFICIENT CAUSE

According to Veda and Sāṅkhya, *Puruṣa* is the efficient cause of this creation. The Vedic seers are unanimous that *Prakṛti* being inertial is quite incapable even to stir not to say of taking to motion by itself. Neither it could do anything nor it is able to perform in any manner. How could it then take to *rajas*, i.e., motion from its state of just being at rest. Veda says that *Puruṣa* is instrumental in causing it to take to motion, thereby initiating the process which culminated in the efficient delivery of the universe through its transformation, a work superbly done. It is *Puruṣa* that determines the process, lays down its rules, activates *Prakṛti* to run through the same to get transformed into a self-contained completely whole system which is procreative by nature and of which we are ourselves a part.

Even Modern Physics with its tremendous development right down from the earth to its billions of light years' reach in space, with unprecedented amount of resources including the entire scientific advancement and the finest brains involved, still continues to fight shy of answering a direct question as to how the inertia-gripped energy could get activated on its own? The entire matter is nothing but a transformation of energy. Einstein has laid down his famous formula which determines the amount of energy involved in such a conversion into matter. The formula has withstood the test of all experimentation and has opened up the highway for scientific development in all the fields, yet the formula works only when the energy is in motion. The question to this day remains unanswered as to how the energy from its state of rest could become active? The motion was and is an effect brought upon the energy by some other cause. Only Veda answers this question most emphatically in a scientific manner. Modern Physics unhesitatingly affirms energy to be the material cause of creation and without doubt holds energy to be inertial by itself.

To any keen observer universe exhibits a number of characteristics and symptoms which are non-material. The entire process of creation runs in a sequence of cause and effect. Each and every manifestation, each occurrence happens to be the effect of some cause which in turn happened to be the effect of an earlier cause, and so on. If there is no cause, there won't be any effect. The matter in universe, through systematic intermediary steps, each accounting as a cause to its succeeding effect, can be traced back to the final material cause, i.e., *Prakṛti*. But what about the non-material effects? The same cannot be traced back to the material cause.. How could they occur and continue to occur or be present?

We find that the transformation of *Prakṛti* in motion is creative. Even destruction results in creation; so it may well be deduced that the motion is purposeful. Its purpose is to achieve some form which we call manifestation. Motion is material but a purposeful motion is material plus. It is not just material as the purpose accompanying the motion can hardly be considered material. Moreover the motion picks up the right direction to its effect which turns in the sequence of cause and effect resulting in manifestation. Agreed that entire *Prakṛti* or energy in motion does not get transformed into matter. Only a small percentage gets manifested into matter. But the thrust of motion is nevertheless creative, which makes it purposeful. And it is along a process which takes it towards that purpose. The process betrays knowledge. *Prakṛti* or energy by itself does not lay any claim to knowledge.

Modern Physics prescribes Newton's three laws of motion. The second of the three laws states force to be the cause of motion. It lays down the formula Force = mass x acceleration. According to Newton, force is an external agent. But external to what? Certainly not to the entire field of energy all around. The electromagnetic force makes its appearance only when energy takes to motion, and not at its state of rest. The gravitational force results from a mass, and the mass results from a process of condensation undergone by energy in motion. Hence, force does not answer the fundamental question as to how energy at rest could get activated on its own. In spite of the first law of motion given by Newton, Modern Physics believes that whatever is in motion was at rest at some point of time. This leads us to conclude that entire energy at some point of time was at rest. At such a state of existence the force could not put up its appearance for want of motion.

THEN HOW DID ENERGY TAKE TO MOTION INITIALLY?

Modern science has no answer. Only Veda has a categorically positive answer which also accounts for the hitherto unexplained non-material characteristics displayed by universe as well. It also satisfactorily answers a few other questions which are still baffling the physicists. The Vedic answer deals with the source of the non-material characteristics manifestly present in creation. Vedic *Prakṛti* or energy of modern science is inertial. It cannot take to motion on its own, and as such cannot transform itself into matter or form. The dominant force cannot appear unless the energy takes to motion. But energy did take to motion whereby the universe came into being. Undoubtedly the motion was generated by a dynamic source independent of energy. Such a source is accountable to all the non-material values manifest in universe, values that are not the products of a transforming energy. The existence of such a being reveals the vista of the entire knowledge and dynamism inspiring all of such characteristics of the system we call universe. Veda names such a being as *Puruṣa*.

Puruṣa has been described in all the four Vedas. *Puruṣa sūkta*, i.e., the wise saying regarding *Puruṣa*, appears in each Veda with some variations in *mantras*. This shows that the subject has been dwelt upon emphasising various aspects of its existence and role in creation. We propose to present below a whole *sūkta* from Yajur-Veda which deals with it. But before that let us examine the noun *Puruṣa* first to see what it denotes. It may be added that this Divine Being has been addressed in Veda in all of the three genders, i.e., masculine, feminine and neuter. But in this text we have preferred to use only neuter. As such even if the original text displays a masculine pronoun its English translation carries its neuter form.

Maharṣi Dayānanda Sarasvatī in his Introduction to the Commentary on Ṛgveda etc. explains the term *Puruṣa* in detail. First we present the same here:

"puruṣam puriśaya ityācakṣīran' (Nir. 1/13) **'puri saṁsāre śete sarvamabhivyāpya varttate sa puruṣaḥ parameśvaraḥ."**

That which lies throughout the cosmos, i.e., exists engulfing and pervading the whole system, is *Puruṣa*, the Supreme Lord.

"puruṣaḥ puriṣādaḥ puriśayaḥ pūrayatervā. Pūrayati antarityantarapuruṣam abhipretya yasmātparaṁ nāparam asti kiñcid yasmāt nāṇīyaḥ na jyāyaḥ asti kiñcit. Vṛkṣa iva stabdho divi tiṣṭhati ekaḥ tenedaṁ pūrṇaṁ puriṣeṇa sarvamityapi nigamo bhavati." (Nir. 2/3)

(puruṣaḥ) puri sarvasmin saṁsāre abhivyāpya sīdati varttate iti, that which exists pervading and engulfing the entire cosmic system; **(pūrayatervā) yaḥ svayaṁ parameśvara idaṁ sarvaṁ jagat sva-sva-rūpeṇa pūrayati vyāpnoti tasmāt sa puruṣaḥ,** that Supreme Lord by itself pervades and engulfs this entire system in motion, each item bearing its own identity, that is why it is called Puruṣa; **(antariti) yo jīvasyāpi antarmadhye abhivyāpya pūrayati tiṣṭhati sa purùṣaḥ,** that which stands also pervading and engulfing even the souls within is puruṣa, **tamantarpuruṣam antaryāminaṁ; parameśvaram abhipretya iyam ṛk pravṛttāsti**: this derivation of *Puruṣa* is based upon root *pṛ* fully pervading. In support of this interpretation of the innermost pervading Supreme Being, Yāska has quoted a *ṛcā*, i.e., a *mantra*:

(yasmātparam........kiñcit) There is no object which may best completely in all respects that Supreme Being called Puruṣa, none earlier or later than It, nothing that equals or betters It. Also there is neither smaller nor bigger object in any manner that ever came into being, nor any that may be happening, nor any, that would ever occur, is conceivable. That motionless and unstirred one, makes all else stir and is itself still. Like what? **(vṛkṣa iva......bhavati')** As a tree stands wearing its branches, leaves, flowers, fruits etc., likewise the Supreme Lord stands still wearing by pervasion the earth, the sun, etc., the entire universe. That which is the only one unparalleled; none is Its similar or dissimilar or else a second Supreme Lord. Because the said Lord by reposing all through and by filling it has made this entire universe full in all respects that is why that Supreme Lord alone is called Puruṣa. This being *mantra* it becomes the final proof and be acknowledged as such.

Having explained the term *Puruṣa* in the context of what is to follow we begin the presentation of *Puruṣa-sūkta* Y.V. chap.31:

sahasraśīrṣā puruṣaḥ sahasrākṣaḥ sahasrapāt
sa bhūmiṁ sarvata spṛṣṭvātyatiṣṭhaddaśāṅgulam

(Y.V. 31/1)

According to Śatapatha **sahasra** means this entire universe. The term **sahasra** also means innumerable. **(sahasraśīrṣā puruṣaḥ sahasrākṣaḥ sahasrapāt)** That within which the innumerable heads, eyes and feet are existing is also called **sahasraśīrṣā, sahasrākṣaḥ** and **sahasrapāt**, as It expresses through the universe. Like space within which all objects exist, but the space by itself is not attached to anyone, **(sa bhūmiṁ sarvata spṛṣṭvā)** that **Puruṣa** is pervading this earth and whatever has come to manifest from all around and in every manner. The noun **'bhūmi'** is an indication of the earth as well as whatever else has come to be. **(atyatiṣṭhaddaśāṅgulam)** The term **daśāṅgulam** denotes the entire cosmos. The noun *aṅgula* literally means a branch or an organ. **Daśāṅgulam** means having ten branches by way of five *tanmātras* and five *mahābhūtas*. The *tanmātras* cover the field of quantum physics and the *mahābhūtas* the field of relativity. The said **Puruṣa atyatiṣṭhat** stands far beyond the field of all creative activity.

Comments: Dr. Stephen Hawking in his book, *A Brief History of Time*, while discussing various models being proposed presently regarding the origin of universe writes about a situation beyond the field of general relativity and quantum physics and suggests the same as the location for such an origin. He calls such an idea as simply a proposal on his part and discusses the merits of such a possibility. In a later chapter we have quoted Dr. Hawking *verbatim*. But here what we would like to underscore is that Veda avers such a position while describing *Puruṣa* in the very first *mantra* and states Its existence to exceed far beyond the said limitations. It enjoys an existence far beyond the bounds of the two states mentioned above. It enjoys an existence from where It is quite capable of energising the energy element of Modern Physics unrestrained by any law of creation. Not only that, Veda affirms that It did initiate the process of the present universe and this is a routine affair with It which It has been effecting since eternity and shall continue eternally, being the eternal instrument of creation as well as dissolution of universes.

So far as the innumerable organs are concerned, describing Puruṣa they denote the organic unity of the entire cosmic system apparently presenting a multifaceted, multi-dimensional appearance. Like the cells or organs of a human body Veda declares the unitary character of universe in spite of its multitudinous appearance.

puruṣa evedaṁ sarvaṁ yadbhūtam yacca bhāvyam
utāmṛtatvasyeśāno yadannenātirohati

(Y.V. 31/2)

(puruṣa evedam .... yacca bhāvyam) That very **Puruṣa**, as qualified by the adjectives above, is instrumental in the creation of all that were ever born, that will be born and by the use of 'ca' all that is existing presently. There is no other Creator. **(utāmṛitatvasyeśānaḥ)** And It alone reigns the realm of immortality. **(yadannenātirohati)** It stands far beyond all that grows or develops.

Comments: All of this that manifested in the past, is getting manifested in the present or will manifest in future be known and realised as having been caused by Puruṣa. The mantra emphasises two of Its fundamental characteristics. First It is the Lord of immortality, i.e., it is the fundamental cause, eternal, beyond mortality; and second It overextends all creation.

etāvānasya mahimāto jyāyāṁśca pūruṣaḥ
pādosya viśvā bhūtāni tripādasyāmṛtaṁ divi

(Y.V.31/3)

(etāvānasya mahimā) The entire universe, covered in time past, present and future, is Its greatness and glory. **(ato jyāyāṁśca pūruṣaḥ)** The greatness and glory of that Puruṣa is greater than this (infinite), as

(pado'sya viśvā bhūtāni) the entire created universe resides but in one of Its steps **(tripādasyāmṛtam divi)** while Its three steps contain the enlightened field of emancipation and immortal energy.

Comments: This *mantra* presents an idea of its immensity by stating that the entire created cosmos is covered under Its single step, while the remaining domain of the immortal element, i.e., *Prakṛti* which too is eternal, is covered under Its three steps. But the use of noun *step* must not be confused with Its walk. The use is simply figurative to illustrate Its immensity, as would be clear from the next *mantra*.

tripādūrdhva udaitpuruṣaḥ pādosyehābhavatpunaḥ
tato viśvaṅ vyakrāmat sāśanānaśane abhi

(Y.V.31/4)

(tripādūrdhva .... punaḥ) This Supreme Lord, the Puruṣa, also exists higher and beyond all of that which has been indicated by the three steps mentioned above and that indicated by one step as well. Its existence remains unattached from all that. The three steps and one step, i.e., the entire range of four steps in which the universe exists, dissolves and re-emerges, i.e., the extent of Prakṛti, are all within that Supreme Being. **(tato viśvaṅ)** The entire cosmos gets manifested in this manner, **(sāśanānaśane)** the one that exits by eating, which is the biological part, and the other non-eating in which the lifeless bodies of matter are included, both of which get created due to Its ability; **(abhi vyakrāmat)** and It is pervading the same from all sides.

Comments: Veda clarifies that the preceding *mantra* must not be taken to suggest any limitations of *Puruṣa* as the same merely forms a part of its expanse which is infinite, far beyond whatever else that may exist.

tato virāḍajāyata virājo adhi pūruṣaḥ
sa jāto atyaricyata paścād bhūmimatho puraḥ

(Y.V. 31/5)

(tato virāḍajāyata) Due to Its instrumentality, *ākāsa*, the field of light and motion, came to be born; **(virājo adhi pūruṣaḥ)** out of the virāṭa, i.e., *ākāśa*, emerged whatever manifested and held by the same, that is the bodies, etc., of all kinds of beings living and non-living. **(sa**

jāto atyaricyata) That Puruṣa, remains unattached, and beyond all that; **(paścād bhūmimatho puraḥ)** afterwards was born bhūmi, the term being indicative of all kinds of celestial bodies having life including the earth and, all of these were created by Puruṣa after prior supportive creation to sustain later creations.

Comments: It caused the manifestation of the immense field of light and motion, the base for all manifestations that were and are to follow. It reveals the plan of manifestation. It says that conditions for sustenance were first created followed by the manifestation that was to be sustained. This character of the creative process is manifest at each step, and discloses planning behind the entire creation. For example, water was created prior to vegetation which preceded herbivorous life.

tasmādyajñātsarvahutaḥ sambhṛtaṁ pṛṣadājyam
paśūṁstāṁścakre vāyavyānāraṇyā grāmyāśca ye
(Y.V. 31/6)

(tasmadyajñāt .... ājyam) According to *Śatapatha yajña* is Viṣṇu. Viṣṇu is that which pervades the entire existence. Hence, *yajña* is a synonym for *Puruṣa* or Its performance here. Moreover every system that helps, promotes or supports creation is a *yajña* in Vedic context and as such represents *Puruṣa*. The *mantra* says that *Puruṣa* was instrumental in the creation of **pṛṣad** and **ājyam**. **Pṛṣad** is symbolic of all sorts of eatables, grains, while **ājyam** is symbolic of all that is used in cooking. The term **sarvahutaḥ** means that the same may be taken by all and applies to both the grain as well as the cookies made from it. **(paśūṁstāṁścakre vāyavyānāraṇyā grāmyāśca ye)** It caused the creation of animals wild and village-dwellers as also those that could take to air, like birds, and small-sized insects, etc.

Comments: The *mantra* mentions the manifestation of winged creatures, domesticated as well as wild animals, insects etc., upon our earth. As stated in the preceding *mantra* the conditions for the sustenance of these creations had already been created by the time these followed. That means the atmosphere, the rains that created seas as well as the forests. Man had arrived and villages had come up where the domesticated animals were reared. The mention of grains covers the state of farming and *ājyam* discloses cooking and frying.

tasmād yajñātsarvahutaṛcaḥ sāmāni jajñire
chandāṁsi jajñire tasmād yajus tasmādajāyata

(Y.V. 31/7)

(tasmād yajñāt sarvahutaḥ) This has been explained in the previous *mantra*. Here it indicates the same meaning save that instead of the *pṛṣat* and *ājyam* here it covers the knowledge contained in the four Vedas named in this *mantra* which was for taking by all persons without any discrimination of any kind based on sex, status, etc.; **(ṛcaḥ ...... ajāyata)** from that **yajña puruṣa** appeared the four Vedas, namely **ṚgVeda**, **SāmaVeda**, **AtharvaVeda** and **YajurVeda**. The term **sarvahutaḥ** is applicable to all the **Vedas**. The use of double verbs **jajñire** and **ajāyat** indicates the Vedas to be the source of multiple knowledge. And the term **tasmāt** indicates the source of the Vedic knowledge to be Puruṣa, the Supreme efficient cause.

Comments: The *mantra* says by name that all of the four Vedas emerged from the *yajña* of creation caused and directed by *Puruṣa*. It has already been stated in this book that Vedas contain the knowledge of all sciences and arts. The ancient scholars believed all such knowledge that they called *naimittika*, i.e., due to some purpose, to have been revealed to mankind through the process of manifestation in the form of Veda. The idea is that knowledge is not a material product. The brain is the fertile ground for knowledge, but the seed or the formula must be sown therein by somebody, normally a *guru* (teacher). The disciple may adopt that seed according to his/her own field, i.e., mind, but the field by itself is not capable of creating the seed and then proceeding to fertilise it. The life of aboriginals upon our earth is a case in point. Anyway, as we are presenting Vedic Physics, it is incumbent upon us to present the Vedic standpoint. It is for the scholars to ponder over how quite a few developments that are simply inconceivable and unimaginable find their specific mention in Veda, which is believed to be the oldest book in human library, from an era when the science of physics was not so developed as we think it to be today. The Veda, i.e., knowledge in toto is manifest in the *yajña puruṣa* called creation. The ṛṣis had its revelation from the same.

tasmādaśvā ajāyanta ye ke cobhayādataḥ
gāvo ha jajñire tasmāt tasmājjātā ajāvayaḥ

(Y.V. 31/8)

(tasmādaśvā ajāyanta) That Puruṣa was instrumental in creating animals like the horses **(ye ke cobhayādataḥ)** which had teeth on both sides. This indicates the strain of such animals like camels and donkeys, etc., which have similar teeth formation. These are village animals as they became domesticated, and their mention here is to indicate such character. **(gāvo hi jajñire tasmāt)** That Puruṣa also caused the creation of cows and all that are covered under the term **gāvaḥ**. **(tasmājjātā ajāvayaḥ)** From the same instrumental cause were born goats and sheep.

Comments: No comments.

taṁ yajñaṁ barhiṣi praukṣan puruṣaṁ jātamagrataḥ
tena devā ayajanta sādhyā ṛṣayaśca ye

(Y.V. 31/9)

(taṁ yajñaṁ .... jātamagrataḥ) That which appeared first of all, the Creator of the cosmos, such **Puruṣa**, and the self-contained whole, *yajñam*, i.e., creation, both deserved, deserve and shall deserve to be **praukṣan** venerated **barhiṣi** within the space of their heart by all. **(tena devā ayajanta sādhyā ṛṣayaśca ye).** The **devāḥ**, scholars, **sādhyāḥ**, the men of wisdom, **ṛṣayaśca ye**, and those who were the seers of the *mantras* as well as other persons appeared.

Comments: The *mantra* states that having realised the fundamental instrumental causc in creation and its manifestation at each and every stage in the process of creation, this cosmos which has been called *yajña-puruṣa* by Veda and the efficient cause of this creation, Puruṣa, ought to be venerated by all persons of wisdom. The realisation of the working of the cosmos blossoms the mental and spiritual capacities of the animal called *Man.* The Veda stresses upon the *yajña* of creation as the manifestation of the *Puruṣa,* the cosmic being. Hence the *yajña-puruṣa* and the *Puruṣa,* i.e., the creation and the Creator both deserve veneration.

yatpuruṣaṁ vyadadhuḥ katidhā vyakalpayan
mukhaṁ kimasyāsīt kiṁ bāhū kimūrū pādā ucyete

(Y.V. 31/10)

(yatpuruṣaṁ vyadadhuḥ) For holding and grasping that Puruṣa **(katidhā vyakalpayan)** in what manner the same may be visualized or conceptualized? Vedas raise such a question **(mukhaṁ kimasyāsīt)** What may be conceived as Its *'mukham',* the face or the best part? **(kiṁ bāhū)** what may be Its arms or the manifestation of Its power and creative capacity? **(kimūrū)** What are Its shanks, that is, the manner of Its support and **(pādā ucyete)** what may be called Its legs, that is, the mode of Its carriage?

Comments: The questions present the base for the personification of *yajña-puruṣa,* i.e., the creation to emphasise the oneness of the whole system. The next *mantra* presents the same.

brāhmaṇosya mukhamāsīd bāhū rājanyaḥ kṛtaḥ
ūrū tadasya yadvaiśyaḥ padbhyāṁ śūdro ajāyata

(Y.V. 31/11)

(brāhmaṇosya mukhamāsīd) Brāhmaṇa means a person well versed in Veda. Meaning of noun Veda is knowledge and the Vedas themselves are the source of all knowledge. Hence a Brāhmaṇa is a source of knowledge. Wherever there is knowledge, whosoever is the possessor of the same, is a Brāhmaṇa and a Brāhmaṇa is the *mukha,* which represents the face, the mind and/or the voice or the best part. Wherever such a state is evident that be considered the manifestation of the wisdom and knowledge aspect of Puruṣa in the creation. **(bāhū rājanyaḥ kṛtaḥ)** Extending the personification of *Puruṣa* in the manifestation of creation, or the society Veda says that the bāhū, which literally means both the arms, but are considered the instruments of valour indicating such beings that are full of valour and strength and may serve as protector, are called *rājanyaḥ.* A *rājanyaḥ* happens to be *bāhūvīryaḥ,* i.e., one excelling in valour/strength/power/energy. The creation of such beings be treated as the *bāhū* or arms of *Puruṣa.* Any system seeks its support from the correct performance and right behaviour of its parts. **(ūrū tadasya yadvaiśyaḥ)** Its constituents, and the constituents of any social order derive its support from the manifestation of such parts and their conduct which denote the ūrū or the shanks in a person and such constituents predominantly seeped in such a behaviour are called *vaiśya* by Veda. **(padbhyām śūdro ajāyata)** Lastly, the term *padbhyām* is derived from the root *pad* meaning *gati,* which indicates that it covers three realms of interpretation, i.e., of knowledge, of motion and of accomplishment. Backed by the three parts mentioned above *padbhyām* or two legs are instrumental in carrying one to the achievement/s within the system or for the entire social order. All those which are deficient in the other three fundamental qualities or characteristics come under this fourth manifestation. They are responsible to keep the system or the social order active towards the achievement of prosperity for the benefit of the creation or the society which reflects the person of Puruṣa in its totality.

It may be added further that this Vedic view can be applied to any social order and the same was applied to the entire human society on this earth. A rāṣṭra commonly understood as a nation in English language, came to be evolved and developed in ancient India during the Vedic era based upon a system known as *Varṇa Vyavasthā* or the social order based on *Varṇa,* i.e., the characteristics as narrated above. It must also be understood that such organ-like placement in a society unified as a person presupposes its conduct in as highly organised a system as the human body. There are no exclusive compartments and all of its cells are nourished through the same blood or progeny rejuvenating the society all the time.

Comments: In the preceding *mantra* Veda of its own proposes a question as to how such a Being be visualised in the created world, and then proceeds to answer in this *mantra.* According to Veda the yajña of creation is observable in each part, and the *yajña puruṣa* may be perceived there. Veda describes such a visualisation in terms of a human form as well as a society. The head being the best part as the repository of knowledge, all wise persons, artists, scholars, scientists, etc., should be visualised as the superior part of the social or national *yajña-puruṣa* and be called *Brāhmaṇa.* All persons of valour be imagined as the arms of that Being and be named *Kṣatriya.* All those devoted to arranging and providing wherewithal for the sustenance of the said social *Puruṣa* be supposed to form Its shanks, the main support, and be known as Vaiśya. Finally all those remaining but not from any of the three specialised conducts as detailed above, ought to be devoted to provide mobility to the whole Being, be conceived as Its legs, and be called Śūdra. Thus the entire human society, or a nation is visualised in this manner as a replica of *yajña-puruṣa* and each performance of each individual of the society done in furtherance of his role is in veneration of the said Puruṣa.

candramā manaso jātaścakṣoḥ sūryo ajāyata
śrotrād vāyuśca prāṇaśca mukhād agnirajāyata

(Y.V. 31/12)

(chandramā manaso jātaḥ) The moon be visualised as a manifestation of Its mental process; **(cakṣoḥ sūryo ajāyata)** the sun appeared from Its illuminated vision; **(śrotrād vāyuśca prāṇaśca)** Its ears symbolising space provided life-bearing atmosphere; **(mukhād agnirajāyata)** the electricity and fire appeared from Its mouth. The personification is extended to the solar system. The moon is known to exert some influence upon the mental life of the beings upon the earth, to which Veḍa is alluding here.

Comments: This again is the extension of personification to our solar system.

nābhyā āsīdantarikṣam śīrṇo dyauḥ samāvartata
padbhyāṁ bhūmirdiśaḥ śrotrāttathā lokān akalpayan

(Y.V. 31/13)

(nābhyā āsīdantarikṣam) Its navel is formed of the space between the sun and the earth or the planets; **(śirṇo dyauḥ samāvartata)** Its head is the sun or the stars that exist; **(padbhyām bhūmiḥ)** the earth or the planetary bodies form Its feet; **(diśaḥ śrotrāt)** from the ears are the directions; **(tathā lokān akalpayan)** likewise all other celestial bodies be conceptualized from It.

Comments: The preceding one and this *mantra* together, extend the personification of the *yajña puruṣa* to the entire cosmos, with our earth-like planets representing Its feet, and the head extending to all the stars, with intermediary locations denoting various body organs. Thus standing at Its feet upon this earth we, or the men of wisdom amongst us, visualize the immense personality of the Creator through Its creation.

It may be further elucidated that though the noun *Puruṣa* conventionally denotes an individual, yet this Vedic Being is utterly formless. Its attributes or characteristics manifest in the universe caused to have been created by It by employing *Prakṛti* which can take material form. But such personification, as stated above, is to impress that the cosmos is one whole orderly personality and not a crowd of innumerable bodies. All sub-systems within the cosmic whole be visualized as organs of a single body system. Similarly it ought to be realized that the human society or the biological world of which the humans are the best known specimen, and are pointed out as such through the above personification by way of forming four important organs, is also a collective whole entity and not a crowd of individuals or the so-called races and nationalities. Like the human body, any thorn pricking or ailment/suffering appearing in any of its parts is a bother for the whole body, similarly the total humanity as a *Puruṣa* ought to take care of and attend to that in case any fault or malady occurs anywhere in the entire system, be it within our reach in the universe, or on our earth or in our societal entity.

The manifestation of Supreme Being, i.e., *Puruṣa* is to conduct the *yajña* of creation at universal level, which is pointed out in the *mantras* that follow:

yatpuruṣeṇa haviṣā devā yajñamatanvata
vasanto'syāsīdājyaṁ grīṣma idhmaḥ śarad haviḥ

(Y.V. 31/14)

(yatpuruṣeṇa .... yajñamataṇvata) The 'yajña' that is manifested by the said Puruṣa was adopted by the learned persons who expanded its performances in various ways. **(vasanto'syāsīdājyam**

.... śarad haviḥ) This part of the *mantra* further illustrates the form of the *yajña* being expanded to be performed by *Puruṣa* round the clock year by year. **Vasanta**, that is, the spring season, is like the clarified butter for oblations for the same; **grīṣma,** the summer season, is the burning fuel and **śarad,** the autumn season; is the material being offered as oblations. The simile is remarkable as the autumn offers the dried up vegetation which serving as fertiliser causes the generation of the clarified butter of green foliage.

Comments: No comments.

saptāsyāsan paridhayastriḥ sapta samidhaḥ kṛtaḥ
devā yadyajñam tanvānā abadhnan puruṣam paśum

(Y.V. 31/15)

(saptāsyāsan paridhayaḥ) Seven borders for this universe were caused to be created—first the earth's border, second inter-planetary space border, third border is the sun, fourth is the inner space within the galaxy, fifth is galaxy border, sixth is the intergallactic space and seventh the world of galaxies—seven borders are around this earth as well like atmosphere, ionosphere, etc; **(triḥ sapta samidhaḥ kṛtaḥ)** together with three times seven, i.e., twentyone *samidhaḥ,* that is, wood fuel : these are one *Prakṛti,* five *tanmātras,* the quantum seeds of *mahābhūtas* each, five *mahābhūtas,* five senses and five organs developed in the biological creation, a total of twentyone. **(devā yadyajñam .... paśum)** The Puruṣa created *yajña puruṣa* bound in its visible form, i.e., the world, by the above-named devas ought to be taken as their ties with Puruṣa through the *yajña.*

Comments: Both the 14th and 15th *mantras* describe the *yajña* initiated and being performed by *Puruṣa* at solar as well as universal levels. In the 15th *mantra* at the end the *Puruṣa* itself has been called *paśum.* Conventionally a *paśu* is an animal. But here it denotes what is visibly manifest. In the preceding *mantra,* i.e., the 14th one, the seasons upon our earth are described as the performance material of the *yajña,* while the same is stated next as being performed at the cosmic level with involvement of twenty-one elements. The seven spheres around our earth and an equal number of borders of expansions within this cosmos, cover the relevant part of the earth dwellers and the whole of cosmic expanse as the indivisible solitary manifest of Puruṣa for physical eyes to see and minds to meditate and realize the same.

yajñena yajñamayajanta devāstāni dharmāṇi prathamānyāsan
te ha nākam mahimānaḥ sacanta yatra pūrve sādhyāḥ santi devāḥ
(Y.V. 31/16)

(devāḥ yajñena yajñam ayajanta) During the summer the intense sunrays form the oblations thereby assimilating the heat of the soil upon the earth through the resultant rains. **(tāni dharmāṇi prathamāni āsan)**. Such effects like rains etc., caused by the intense heat of the sunrays are priority based or protected, as the same result from a long spell of intense sun heat. **(ha te mahimānaḥ devāḥ nākaṁ sacanta)** And after the summer season those intense rays get transferred through the very sun **(yatra pūrve sādhyāḥ santi)** where earlier rays had their existence, i.e., the atmosphere **sādhyāḥ** = **devāḥ** = rays of the sun.

Comments: It may be noted that the last part of the *'mantra'* further states a scientific fact by way of a corollary. The transfer of heatened rays in fact is communicating a heat transfer from the crust of the earth to its atmosphere due to rains. The rains caused by sun did not annihilate heat within the affected surface. The same got transferred to water which again rose to the atmosphere wherefrom it had earlier come down.

adbhyaḥ sambhṛtaḥ pṛthivyai rasācca viśvakarmaṇaḥ samavarttatāgre
tasya tvaṣṭā vidadhadrūpameti tanmartyasya devatvamājānamagre
(Y.V. 31/17)

(adbhyaḥ .... samavarttatāgre) For the creation of **pṛthivī**, that is, the causative of molecule formation, and thence forth the entire material structure of the universe symbolised in the term **pṛthivī**, that **Puruṣa** caused atoms created by **mahābhūta jala** to perform.This last step was effected out of a succession of cause-effects, namely, atoms at the *jala* effect were caused to be created by the cause of **agni mahābhūta**, **agni** was effected from **vāyu**, **vāyu** from **ākāśa**, and **ākāśa** from the causeless **Prakṛti**. Before that, this entire design of creation was held in the state of causation within the One that caused all of this to be created, namely, **Viśvakarmā** or the creator of the universe. The details stated here from the last state of **pṛthivī** backwards are covered by the single term **pūrve** meaning earlier process in the *mantra.* The name **Viśvakarmā** is a synonym to **Puruṣa (tasya tvaṣṭā vidadhadrūpameti).** Its creator imparted the universe its **rūpa**, i.e., the material appearance. The term **tvaṣṭā** conventionally means a carpenter. The one which causes the shape of universe is its carpenter, another synonym of that Supreme Being. **(tanmartyasya devatvamājānamagre)** That the achievement of that wisdom is the wholesome duty of the mortal man has been made known to all in advance through Veda.

Comments: In this *mantra* Puruṣa has been described as Viśvakarmā, the Creator of the cosmos or of the total manifestations. The term *viśva* means the cosmos; it also covers the totality; and the other term *karmā* means the doer or creator. It has also been mentioned as Tvaṣṭā, the carpenter. A carpenter takes a log and transforms the same into a number of articles different in shape, measure, weight, etc. All articles made out are different yet they are all made of wood, which they ultimately remain. The instrumentality of Puruṣa causes effects through the inertial Prakṛti, i.e., energy, like a carpenter to a log. Its efficiency causes transformation of the inertial material to numerous shapes and forms and yet the entire lot of creation remains nothing but energy in itself. But in this case the log appears to be working upon itself for production. The direction given to mankind in the last quarter is particularly noteworthy. It says that the wisdom of this carpentry, i.e., the science of creation and the craft of its application be learnt by man as a duty. This is being made known in advance through Veda to all beings.

vedāhametaṁ puruṣaṁ mahāntamādityavarṇaṁ tamasaḥ parastāt
tameva viditvātimṛtyumeti nānyaḥ panthā vidyate ayanāya

(Y.V. 31/18)

(vedāḥ......parastāt) Knowing what one would become a man of knowledge? The seer of the *mantra* replies:When that one imbued with all of such symptoms, attributes, characteristics, etc., as mentioned earlier, one who is the greatest among whatever exits, one who is illuminated because of being the embodiment of all sciences and whose existence is beyond any darkness born of any kind of ignorance and unwisdom, such a Supreme Being, Puruṣa, becomes known to me, I know for sure, I am a man of knowledge. **(tameva viditvātimṛtyumeti)** A man only by knowing such a Puruṣa, the Supreme Being, steps beyond death, achieves emancipation. **(nānyaḥ panthā vidyate ayanāya)** No other way exists for the achievement of the total welfare.

Comments: The ṛṣi says that he has come to know and realise Puruṣa which is the greatest of the greatests. He further says that by knowing the same mortality may be overstepped assuredly. There is no other way to achieve such a state.

prajāpatiścarati garbhe antarjāyamāno bahudhā vi jāyate
tasya yonim pari paśyanti dhīrāstasmin ha tasthurbhuvanāni viśvā

(Y.V. 31/19)

(prajāpatiś.......jāyate) It alone is the Lord of all that exists, and is eternally present by pervading the innermost and outermost of the biological and material existence, and is Itself birthless. Due to its capability the entire universe gets born in many ways and forms. **(tasya yonim pari paśyanti dhīrāḥ)** Its appearance is observed all around by men of patience. **(tasmin.....viśvā)** Within It exist all of the galaxies and the entire universe.

Comments: No comments.

yo devebhya ātapati yo devānāṁ purohitaḥ
pūrvo yo devebhyo jāto namo rucāya brāhmaye

(Y.V. 31/20)

(yo devebhya..........purohitaḥ) The Supreme Being that enlightens the learned persons with knowledge, that which also takes care of their wellbeing in all respects, **(pūrvo ..... jāto)** which is by Itself the initial and topmost among the learned beings, **(namo .... brāhmaye)** we bow to that Being which can make truth our taste.

Comments: The preceding *mantra* describes Its existence, capacity and capability all infinite. This *mantra* is an expression of veneration born out of the realisation of Supreme Being presiding over the cosmic activity.

rucam brāhmam janayanto devā agre tadabruvan
yastvaivam brāhmaṇo vidyāttasya devā asanvaśe

(Y.V. 31/21)

(rucam..........devā) The learned persons keep learning and earning the knowledge of the Supreme Being **(agre tadabruvan)** and narrate or deliver the same onwards. **(yastvaivam.....vidyāt)** The person who knows it fully in this manner **(tasya devā asanvaśe)** his pleasure-seeking senses and organs remain under his discipline or control.

Comments: The *'mantra'* says that it is expected of those who are men of learning to keep on earning more and more knowledge of the Supreme Being from its manifest the cosmos and the creative activity running through the cosmos. They are also expected not to indulge their pleasure-seeking tendencies in trying to seek satisfaction of carnal desires or trades, but rather set their example to disburse the knowledge so gained to better life and conditions all around.

śrīśca te lakṣmīśca patnyāvahorātre pārśve nakṣatrāṇi rūpamaśvinau vyāttam
iṣṇanniṣāṇāmum ma iṣāṇa sarvalokaṁ ma iṣāṇa

(Y.V. chap. 31 ends)

In this final *mantra* of the *Puruṣa sūkta* in Y.V., the biggest *sūkta* in all the four Vedas, the subject matter is concluded by returning to personification to impress upon the oneness of the system created followed by a prayer or an auto-suggestion. **(śrīśca......patnyau)** The *mantra* says that the splendour and prosperity are your (of Puruṣa) two companions in this universal *yajña.* The term used is **patnī** which conventionally denotes a wife. But the Pāṇini grammar specifically rules that the term has to be coined to denote a female associate in the performance of a *yajña.* That is why the English term companion has been used. The idea is that the two keep company. It is the *yajña* being performed by *Puruṣa,* **(ahorātre pārśve)** like day and night, or the sun and the moon, as two sides. The stars and constellations are your image, **(aśvinau vyāttam)** that is, the earth and the world of light, the space between the two is like your mouth. Now follow the prayers **(iṣṇan iṣāṇa amum)** O Puruṣa, the emancipation is at your wish, **(ma iṣāna)** please grant us the same/ **(sarvalokam ma iṣāṇa)** grant us all the worlds. The last prayer implies Śrī and Lakṣmī, and seeks that the same be bestowed upon us.

Comments: The terms Śrī and Lakṣmī respectively, denote creative wisdom and material wealth. The ṛṣi prays that Puruṣa may grant both to us. Our dictionary does agree with us in this interpretation as far as it states Śrī to denote intellect while Lakṣmī to represent wealth and riches. According to the Vedic thought as explained earlier the two have been described as patnī of Puruṣa in this last *mantra,* a term which traditionally means a wife, but has been interpreted by us as a companion. The incorporeal *Puruṣa* has Its wisdom and capability to direct *Prakṛti* to perform to create the material cosmos. As such both wisdom and capability are Its eternal companions. If the material manifestation is taken to mean a transformed *Prakṛti,* that, too, being unborn like *Puruṣa* stands as an eternal companion of the latter in all times. The root *lakṣ* from which the latter is derived means to view, to mark. Both meanings are based upon visibility. The Lakṣmī has to be visible which even the otherwise invisible energy becomes when transformed into matter.

Special Comments on the whole sūkta—

It may be mentioned after the conclusion of the above *Puruṣa sūkta,* that *Puruṣa* is the most important factor in Vedic Physics. It is the only infinite Being that the Veda knows and acknowledges. *Prakṛti* is by itself finite in expansion though the same conserves and enjoys eternal existence. But *Puruṣa,* being finer, pervades *Prakṛti* in its entirety, and extends by Itself far beyond the realm of both the quantum physics and the field of relativity as stated in the initial *mantra* of the *sūkta* above in the words **atyatiṣṭhaddaśāṅgulam** already explained there. Not only that, as soon as Veda in the third *mantra* states that all the manifested bodies in the universe fall within one step of the said *Puruṣa,* while the unmanifested formless *Prakṛti* extends to the three steps, the very next *mantra* scrupulously proceeds to point out that the extent of *Puruṣa* is not confined to the said 3+1 steps, but exceeds far beyond. The Vedic exposition of this aspect of *Puruṣa* makes two aspects clear: that the field of *Prakṛti* ends somewhere within the infinite and eternal existence of *Puruṣa*, and that *Puruṣa* stands or exists far beyond the entire realm of the cosmic universe including both the manifested and unmanifested extents. Such an infinite existence of Puruṣa, as we shall later on see, will solve another riddle of Modern Physics, the one raised in the beginning of this chapter, as to how inertial *Prakṛti* (energy to the physicists) came to take to motion to initiate the process of creation.

The Veda has used the term *Puruṣa* to convey a lot many meanings both directly and in an implied manner. The Veda wants a *puruṣa* to know and realise the *Puruṣa* and Its work of creation which is termed as *yajña*. A *puruṣa* is each individual human being. The *Puruṣa* is the causeless, birthless, eternal existence of wisdom and dynamism more fully explained in a following chapter entitled Supreme Master of Sciences. The work done by the *Puruṣa* is the creation of universe by energising *Prakṛti* through permeation of Its own personality of wisdom in it. Such an action needs no physical activity as the *Puruṣa* enjoys Its existence pervading and transcending the realm of inertial *Prakṛti*. The existence or the state of Being engulfs all the states from zero, that is, a point to infinity in mathematical terms leaving no exception whatsoever. That is why the entire activity of creation is Its own play willed by It for Its own self. Hence each and every activity or motion in the universe effected through the inertial *Prakṛti* remains complementary towards the whole system of manifestation called universe which ticks with wisdom of Puruṣa. As such Veda has named the creative activity a *yajña* and the universe a *yajña puruṣa* reflecting the personality of the Absolute.

Let us recapitulate the *Sūkta* explained above. The first *mantra* introduces us to that Absolute Being and Its reach, while the second *mantra* to Its work covering whatever was, is, and is to be. The third *mantra* conveys the extent of *Prakṛti* and its creation within the *Puruṣa*, and the fourth one restresses Its Being. The fifth *mantra* states the cosmos with the plan of its manifestation, i.e., the grounds for sustaining are caused to emerge first to be followed by whatever is to be sustained. A passing reference to material creation of cosmos has also been made.

The sixth *mantra* opens up the creation of the biological world. The use of noun *yajña* for the creative activity is also introduced which is carried forward in following *mantras.* The seventh *mantra* speaks of the manifestation of the knowledge and wisdom of the *Puruṣa* in the said *yajña* of creation and its emergence by way of four Vedas. The eighth *mantra* completes the brief reference to the animal world. The ninth *mantra* stresses the importance of the wisdom contained in Vedas and its value for human beings.

From the 10th *mantra* begins the presentation of the personification of the *Puruṣa* to be viewed through Its work, the creation, by raising questions related to a human form, i.e., an individual *Puruṣa.* The following 11th *mantra* presents answers to the said questions thereby completing an imaginary form of the *Puruṣa* in a social or national set-up or even involving the entire human race. In the next four *mantras* the personification has been extended to the solar system as well as to the universe. The 16th and 17th *mantras* emphasise knowing the *Puruṣa* through the *Yajña Puruṣa,* that is the universe. The 18th *mantra* lays final stress on such a knowledge of the *Puruṣa.* The 19th *mantra* stresses patience for gaining such a vision of the incorporeal *Puruṣa* pervading the *Yajña Puruṣa* of the creation. The 20th *mantra* is an expression of reverence for the same followed by the penultimate *mantra* communicating that with such a knowledge an individual becomes a master of his self. The last *mantra* contains a prayer for the entire wisdom and entire prosperity in the universe being granted to us.

In short this *sūkta* presents the central thrust of the Vedic knowledge that the universe is a manifestation reflecting the personality, attributes, wisdom, etc., of the incorporeal *Puruṣa* through the innumerable multiplicity as well as diversity converging on and serving the singular paramount Supreme aliveness of eternal existence so that human beings may learn Its science from Veda and Its creation and emulate Its *yajña* in their own individual, social, national and universal life for all times to come. The purpose of Vedic Science of Physics is to demonstrate that each point opens upto infinity.

— 17 —

YAJÑA

Can the central thrust of the entire Vedic thought and knowledge be summed up in a single word which, like a point expanding itself in concentric circles, possesses the wholesome capability to cover infinity in all dimensions including time? The answer to such a question is a most emphatic and most positive Yes. That one word is *Yajña*. No language possesses a synonym for the same. The Sanskrit language, more specifically the Vedic language, has given quite a few words which, like banyan seeds, contain a unique tree of knowledge and wisdom. *Yajña* is one of the most potent ones among them.

In English language this word has been stated to mean a sacrificial fire, an act of offering oblations, sacrifice., etc. which hardly explain its content. A *Yajña* extends from an individual to the universal level. In fact the creative activity which blossoms into the form of a universe is the performance of a *yajña* itself. All science, all knowledge, and all creative activity are covered under this one noun. It is no ritual. It is the very essence of all constructive and positively beneficial thoughts and actions, nay, the very thrust and direction of nature as a whole towards sublimation.

As the subject of this book is Vedic Physics, this chapter on *yajña* is being included solely with the intention of presenting an idea of this central thrust of the Vedic thought behind all scientific thinking in general and this science in particular. As usual we examine *yajña* etymologically to unfold its contents.

Derived from the root *yaj devapūjāsaṅgatikaraṇadāneṣu* with the addition of suffix *na* the noun *yajña* basically presents threefold meaning, namely, *devapūjā*, *saṅgatikaraṇa* and *dāna*. Let us take them one by one:

Devapūjā: This is a compound of two words, namely *deva* and *pūjā*. The noun *deva* is derived from the root *div* which makes triple appearance in Sanskrit grammar. The first one is stated to denote 10 meanings among which the last one being *gati* denotes three domains as has been pointed out earlier at various places. This increases the total of meanings to 12. The remaining two roots are stated to denote single meaning each. Thus the root *div* covers 14 fields of meaning. For our limited purpose we propose to discuss only two meanings stated against the first root *div*, namely, *dyuti* and *gati*. *Dyuti* is luminescence and *gati* denotes knowledge, motion and accomplishment.

The noun *deva* derived from the root *div* to denote luminescence covers any object or material that emits or reflects light. The entire universe gets manifested in the field of light. The whole lot of stellar systems is emitting light, while particles, atoms and their combinations discharge the same, and all material bodies in general reflect light. Even the so called dark matter gives out rays. The only exception may be a black hole, but its peculiarity is that it does not let light escape. So it is also not an exception in the scientific sense, as the generation and reflection of light does occur, but the same gets arrested at its source.

In short all matter in whatever form or state consisting of the universe collectively and severally, is covered under the term *deva*. Further the light being energy, all forms of energy as well come under this label. Even *ākāśa*, the field of light, gets covered under it as has already been explained earlier through the interpretation of the epithet *mahādeva*, the great illuminated one, in chapter 11.

The second word *pūjā* of the compound means veneration or reverence generated on account of knowledge. Thus the complete term *devapūjā* denotes a spirit of veneration or reverence born out of the knowledge regarding universe as a whole and the various systems and bodies forming the same. As true knowledge is simply scientific knowledge, the first realm of the meaning denoted by root *yaj* not only points to but rather prompts a scientific study of the creative process in all of its ramifications and places such a study on the highest pedestal in life by giving it the name of a *yajña*. No word is more sacrosant than that. This aspect covers all the physical sciences.

In nearer terms parents, teachers, all learned persons like ṛṣis and scholars, all artists and craftsmen, in short, any person or being who is instrumental or guide in any pursuit for knowledge in any field whatsoever, is a *deva* and as such an object of veneration or reverence. In the ancient Indian tradition there are references of animals, birds, even insects being treated as *gurus* (teachers).

At this juncture, it appears pertinent to point out that worshipping an idol or any other object as a ritual is no *pūjā* in the true sense denoted by that term. It is the knowledge which is the real object to be venerated and all objects can at best be a source or an instrument for the revelation of knowledge. The reverence is the spirit in which the knowledge and the source of it should be treated. There is absolutely no blind approach. With open eyes, an alert mind and the rest of the senses alert, any milking of knowledge is to be performed with a temperament oozing respect.

To be more specific, it is the *Puruṣa* factor, the last but not the least element of creation, enumerated by Sāṅkhya, towards which the entire thrust of the term *devapūjā* acts as the pointer. According to both Veda and modern science, *Prakṛti* or energy being inertial is utterly incapable by itself of acting not to say of acting in a superbly intelligent and orderly manner. Modern science is mute, but Veda declares *Puruṣa*, as the efficient cause of creation, to be the source of all intelligence and knowledge behind the concept of universe, and the source of thrust to *Prakṛti* to execute the same. The knowledge behind the manifestation of universe at each and every stage is the knowledge emanating from *Puruṣa*. As and when such a realisation dawns upon the human mind that inert *Prakṛti* is dancing to the tune of *Puruṣa* by way of manifestation of itself in a universe, the possessor of that mind realises the fundamental unity of the universe and the eminence of *Puruṣa* pervading it.

Saṅgatikaraṇa: The word *gati* means motion and *karaṇa* is doing or creating. Thus *gatikaraṇa* means creation of motion or motions. The addition of prefix *sam* changes the whole term to mean unification of two or more motions into a system. The other two meanings, knowledge and accomplishment apart from motion already stated, are *ipso facto* involved. Creation of such a unified system is executed on the basis of prior knowledge and creation is effected to accomplish the desired object.

The entire universe is a *saṅgatikaraṇa* or a system functioning through the unification of several live multi-stage systems, big and small, each ticking on the basis of a unified order established among the numerous motions within. Right from the micro level Hydrogen atom to the galaxies having diameters beyond a hundred thousand light years this *saṅgatikaraṇa* is effective in a harmonious order. Each atom is a superb example of the same. In bigger bodies the unification of various motions of the planets with the sun on the one hand and their respective satellites on the other is responsible for holding our solar system and all other stellar systems functioning in harmony. All motions are related and integrated so well that one hardly notices such a unification. The famous theory of relativity for which the modern world of physics gives credit to Albert Einstein is simply the cognizance of *saṅgatikaraṇa* by that renowned scientist. Both the general and special theories regarding relativity propounded by him are corollaries of such an integration called *saṅgatikaraṇa*. All astronomical calculations are done taking this *saṅgatikaraṇa* into account.

In the biological world *saṅgatikaraṇa* is ingrained everywhere in each unit from the smallest bacteria to the largest ones that ever roamed upon the earth. One may observe the same prevalent in vegetation no less effectively. The seasons, the daily sunrise and sunset, the rotation of the earth, in short all activities and motions are occurring in a synchronised manner. Even the solidification, liquification and gasification of matter is governed in the same manner.

In human life all innovations as well as what we consider as inventions, all technology and mechanics, all arts and crafts come under this term *saṅgatikaraṇa*. On social and political planes, the cultures and civilizations, the nations and states, the constitutions and polity, inclusive of the prowess in warfare and militarisation, the will to live and die for a value-based life, even the gregarious instinct exhibited in animal life, are manifestations of *saṅgatibhāva* or *saṅgatikaraṇa*. The two Sanskrit terms used here are two sides of the same coin. The word *saṅgati* has already been explained above. The next term *bhāva* denotes being and *karaṇa* denotes doing.

It also includes investigation of *saṅgatikaraṇa* in matter and nature and innovatively applying or employing the same in life. All *saṅgatikaraṇa* done in the spirit of pūjā (reverence) for some universal, selfless cause makes life divine. The contrary may demonise the same.

Dāna: The word *dāna* literally means donation. But *dāna* is not simply donation of money or wealth or giving alms. *Yajña* inspires one to know one's being and donate the same in the spirit of dedication for sublime objective.

The fundamental question related to spiritual investigation has been: Who am I? From the Vedic point of view, and more particularly in the context of *yajña*, the same is no less physical, for the Vedic spiritualism begins at physical level. Physical existence is no less important, and in no case despicable. To understand donation in the context of a *yajña* let us begin with the physical existence.

In physical terms every I is a bundle of matter and energy and the matter and energy of the so called bundle, that is, I is conserved. It does not decay. It is immortal, and cannot be destroyed in any manner whatsoever. And yet we experience gain or loss in our physical existence. The investigation reveals that both gain and loss are confined to the body taken in isolation and is a transitory phase of the process of creation. In reality there is neither any gain nor any loss in the total mass. It is all matter and energy. The I is simply the ego generated as a consequence of life. It has little independent value. But the feeling of a separate system coming into existence associated with a nervous system makes it one of the most potent governing influence upon each individual life. The reality becomes virtually incomprehensible. Relativity changes the perception of reality altogether.

It is the knowledge of science in association with *yoga* prescribed by Veda, expounded by the *ṛṣis* and compounded as a system of self-concentration and deep meditation of one's entire mental energy upon Puruṣa that one realises the reality of the immortality of one's existence and total unity with the cosmos. Such a realisation of science or knowledge of existence changes the perception of the individual self to that of universal self. Such an individual realises that the sole purpose of his being is to support and further the process of creation and the values connected with the same. His life becomes a total dedication for such a cause. The universal personality gets reflected in such a being and his actions. His whole being becomes a person donated to the universe and the cause of creation.

Such a donated life becomes an embodiment of *yajña*. There is hardly any feeling of the individual self; the ego gets completely sublimated in cosmic status of Puruṣa. All of his thoughts and actions, after the attainment of such a realisation of the scientific reality, are in the service of and for the benefit of universe. He feels and acts for others almost like a mother for her offspring. Such a dedicated life is the life donated—in other words a life performing a *yajña*, almost in a manner Veda declares the *Puruṣa* performing the *yajña* of creation.

There is a śloka in Sanskrit which appears worth quoting:

ayaṁ nijaḥ paro veti gaṇanā laghucetasām
udāracaritānāṁ tu vasudhaiva kuṭumbakam

It says: This is mine or somebody else's—such a consideration is done by persons of limited perception. For those who are of generous character the entire world is their family.

A typical anecdote mentioned by Maharṣi Vedavyāsa in his great epic Mahābhārata is worth mentioning to illustrate unimportance of any ritual or the so-called sacrificial fire in the performance of a *yajña*. While describing the performance of a great *yajña* by king Yudhiṣṭhira after ascending the throne, the anecdote is narrated by way of a contrast to the royal performance.

The famous *rājasūya yajña* performed by king Yudhiṣṭhira was a grand show. Thousands of people participated including a number of ruling princes and great scholars. All of the participants were treated most courteously and with respect. A lot of amount in cash and by way of gifts was donated to the priests and more was distributed to the poor. The performance was done under the overall supervision of Lord Kṛṣṇa. After the *yajña* a feast was arranged which concluded with the washing of their hands by the participants who were served water by the members of the king's family. The ground became wet with such a wash. The gathering dispersed.

Then appeared a mongoose from somewhere. The people observed that half of his body was golden. He went straight to the wet piece of earth upon which hands had been washed after the meals earlier, and started stretching his body and rubbing the same upon it. He indulged in such an activity for some time, and retired as if dejected. The onlookers were astonished. Thinking the mongoose to be somebody special, they asked him as to who he was and what he was doing.

The mongoose spoke in human voice and told them that years earlier during a famine he happened to cool his body by wetting the same on a piece of ground rendered wet with the washing of hands after meal at the conclusion of a *yajña*. The part of his body that came in touch with the water spread over there became golden. The water was too little to wet his whole body. Since then he had been going to many prominent *yajñas* and rubbing his body upon the ground rendered wet after the meals, but in vain. The king's *yajña* was a very great *yajña* and he thought that it would do the trick. But alas !

The listeners became highly curious and requested the mongoose to tell them what sort of *yajña* that was. The mongoose told them that during the period of a great famine, a *brāhmaṇa* family of four persons, the *brāhmaṇa*, his wife, his son and his daughter-in-law, could not get even a morsel to eat for days together. Finally the *brāhmaṇa* could obtain four breads and a little rice and vegetable and returned to his cottage with the same. His wife made four equal parts of the food package and distributed them among themselves. They were about to begin their meal when a knock came from outside. The brāhmaṇa arose and came out to find another starving person begging for something to eat. He made the new-comer wash his hands and be seated for the meal.

Returning inside he found the rest of the family waiting for him. They were told of the new comer, and the *brāhmaṇa* picking up his share of food went out and served the same to the guest who ate it. Finding him still hungry the house-owner came in where his wife gave him her share of food for the guest. The stranger consumed the same and still reported hungry. One by one the son and then the daughter-in-law also offered their shares of food which was served to and eaten up by the guest who after washing his hands departed blessing his host. The four died of starvation. It was in that wash water that the mongoose happened to rub his body—people were told, and the mongoose departed.

There was no sacrificial fire, no oblations, no recitation of *mantras* or any other ritual, neither any priest, and yet the author of the epic *Mahābhārata* and himself a great authority on Veda, better known by his title Vedavyāsa to this day, has preferred to include such an anecdote at the conclusion of a famous Vedic *yajña* performed in royal splendour in the presence of many Vedic scholars of the day attending and participating in its performance.

Veda or *yajña* do not preach self-abnegation. They advise sublimation of the self to include all. The *brāhmaṇa* family of the anecdote did not abnegate itself and commit suicide. The self of each of the members was so large as to feel the hunger of a stranger no less acutely and the donation of their food by each one of them was willing and voluntary like a loving mother under scarcity giving her own morsel to her offspring and enjoying the pleasure of giving comfort and satisfaction to her kid over the pain of her own hunger.

There is a world of difference between a *yajña* and an *agnihotra*. The latter literally means a formal performance in which oblations are made to *agni* (fire). But even that is done for the benefit of all. We are not pointing out the *mantras* recited or prayers made. We are simply explaining the oblation to a fire. It is a scientific fact that the contents of anything burnt in a fire get broken up into molecules which riding over the heatened lighter air spread in the atmosphere covering the area and its inhabitants coming under their reach. The material that is used for making oblations is of medicinal value or purifier. It is invigorating and health-giving and acts as a purifier of air and water. In the Vedic science of *Āyurveda*, for healthy living, one of the established modes of treatment is to ask the patient to perform *agnihotra* by making oblations of certain dried up medicinal plants. That is why any translation like '*a sacrificial fire*' is not only a misnomer but utterly misleading. On the contrary it is a wholly scientifically useful exercise.

But a *yajña* basically covers one's attitude towards the world. Any act done with the intention and knowledge for the benefit of a being beyond the realm of one's individual self and for no gain may well be considered a *yajña*. In short, any act done to support and encourage the process of life or creation without any reservations or selfish motives is *yajña*.

An *agnihotra* may or may not accompany a *yajña*. But it is symbolic of the spirit of a *yajña* inasmuch as the oblation made is disbursed all over by *agni* (fire) without retaining even an iota for the performer who, too, may share the benefits with others of the improved atmosphere. It is a replica of the primordial fireball, or the *Hiraṇyagarbha* disbursing its contents for the creative activity. In fact the performer makes such a declaration while presenting an oblation to fire. For example, he says, *Sūryāya svāhā*. This is for the sun, i.e. weather; and then adds *Idanna mama*, It is not for me. *Indrāya svāhā* this is for the Indra, i.e., rain, and repeats, *Idanna mama*, It is not for me. The oblations are made for no individual selfish purpose by the performer but for the system which helps all.

Let us see the *yajña* in our part of the universe. Our sun is burning itself disbursing its heat, light and energy all around as its contribution towards the maintenance of the universal system. Our earth is giving all of its possessions like light, wind, water, space, products, etc. free, for the benefit of all the earthlings. The vegetation is not keeping even an iota of its product for itself. It is available for all who can take it and consume the same; otherwise it is for being scattered to increase production. Even in our bodies the limbs are all working ungrudgingly for the maintenance of the body system. Our heart retains no part of the blood for itself. On the contrary it keeps on constantly pumping the same to the last iota of its energy to make it reach all over the body nursing and repairing the cells.

Even the animal and the insect population is busy playing its role. Like an earthworm cleansing and improving the quality of soil all the time, each one is doing its bit to live in tune with nature, save the most intelligent, most competent and most resourceful creature, the Man. He has been for ages more busy in plundering the natural resources for his personal aggrandizement. Instead of unifying and harmonizing the human society with nature, he has abused his knowledge and skill in causing and organising more and more heterogeneous divisions and precipitating chaotic conditions. The lust for power, money or influence is the rule under one garb or another.

It is noteworthy that Veda declares and emphasises the unity of universe. The so-called diversity in nature is no diversity as such. It is the fundamental unity expressing itself in diverse ways. Like the radii of a circle, each one being different from the rest, even though they all originate from the same point called their centre and basically extend the reach of the point to infinity in concentric circles the circumferences of all of which they continue to hold together as the expanded embodiment of the point of their origin. There is no diversity; it is simply the diverse appearance of the fundamental unity.

If the universe is one comprehensive system as stated by Veda and further determined by the modern science, no action can be performed in isolation. It is simply not possible. Each and every action, even every stirring or motion will deliver its reaction, howsoever subtle, upon the universe like the force of gravitation. The accumulation of unhealthy actions not in tune with the creative process will develop an unhelpful entropy exerting its pull that may hinder or even create chaotic situations in biological or human worlds.

Nature has not created any disorder. All of the disorders having risen in the human society of our times have been created by the actions and deeds of its individuals in pursuit of the satisfaction of their individual or narrow selfish ends. It has led to the abuse of knowledge and the abusive employment of the resources and instruments generated on the basis of knowledge gained.

That is why the Vedic wisdom differentiates between knowledge acquired and knowledge realised. The knowledge acquired gives the person power which tends to feed his ego by going into his head, while the knowledge realised disciplines the conduct of its subject. It also provides its beneficiary a universal vision for the application of the knowledge realised.

In Sanskrit language *sākṣarāḥ* means literates, those who are acquainted with letters. There is a famous saying which says that a reversal of *sākṣarāḥ* is *rākṣasāḥ* that is the demons. A person who has realised the knowledge becomes a living embodiment of that knowledge and we must remember that all knowledge is universal. The realisation of knowledge helps you differentiate between *sat*, harmonious existence or creation and *asat*, heterogeneous to creation, and also provides the ingenuity to employ it only in favour of the former and against the latter.

The present growth and development of scientific knowledge is bringing about its realisation although slowly yet somewhat effectively. Tolerance for a view uncharacteristic of one's own is gaining ground. The consideration of human rights, freedom of speech and expression and what is generally known as democratic way of life is deepening its roots and the growth of the manifestation of the concept of coming together under a universal system, say United Nations Organisation, is in fact a positive step towards inculcating the spirit of *yajña* in the field of polity.

— 18 —

SUPREME MASTER OF SCIENCES

It has been already pointed out that both Veda as well as Modern Physics are unanimous that energy is the material cause of universe. The Vedic noun for such a cause is *Prakṛti* as has been stated in chapter 1. But Veda does state the existence of an efficient cause of creation apart from the material cause, which makes the said material cause deliver the goods. By tradition such a line of thinking takes us to the acceptance of a God who by means of His supernatural powers did the job in a twinkle according to some, in a period of six days according to others or in an otherwise specified or unspecified time as per still other believers. Such a God fails to pass the test of science. That is probably the supreme hurdle which science has not been able to surmount. Hence it continues its search to find some clue that may let it out of this blind alley of causing energy to move the scientific way.

In the chapter on *Puruṣa* the Vedic stand specified by Sāṅkhya regarding this issue has been presented from Veda. Therein Veda summarily states the manner, the efficient cause has been instrumental in the manifestation of a well ordered system called universe to emerge. The point is once again being raked here to examine the same from the angle of modern science. The issue to consider and, if possible, to determine, is why should science concede the existence of a Supreme Being as a scientific fact independent of inertial energy?

It is obvious the riddle cannot be solved by a personal God. Scientific research has proceeded far too long, and its results are by far too weighty, either to be ignored or to bow down to a personal God. To any scientific analysis the very concept of a personal God is a contradiction in terms. It fails at the preliminary enquiry itself. For the last couple of centuries or so, science is closely examining the order of creation, and while following such a track it has grown its stature not only to assume the centre of stage position but even to outshine all other fields of research. Science is trying its very best to determine truth. All along in pursuit of truth its researchers have dared the so called religious belief and even suffered for the same. But the power their work brought to them, forced the Church as well as the world to hold them in esteem and enjoy the benefits, the application of their research provided to the modern world. Thanks to science, Man is taking a close look at the immensity of the creation both at the macro as well as the micro levels.

Before we turn towards Veda to see what it has to say, let us proceed with our enquiry a bit further along the modern scientific course. Such a suggestion must not be misinterpreted to mean that the Vedic way is something other than the scientific approach. Any method which allows freedom for the application of both the processes of analysis and synthesis in total objectivity for the determination of truth must lead to the core or the heart of the riddle. But as has been pointed out earlier modern science is attempting to track the centre from the circumference, while Veda moves quite the other way from the centre towards the circumference. As such there is no reason why the track adopted by science may not lead us to the centre and meet Veda there. If the direction of the search or research is right, the two tracks must open in each other's arms.

So we take the scientific track and proceed. The science of physics has determined that matter is a transformation of energy. Naturally the limits of cosmos must be within the limits of energy. Let us, therefore, see if we observe some characteristics in universe which do not tie up with the nature of energy. Do we observe any features which do not conform to the basic character of energy? If so what does science has to say about them?

The first and foremost is the appearance of motion. The entire universe is constantly on the whirl. But the basic character of energy is inertial. It cannot stir on its own. If motion is a must, as it is, then how could energy take to motion? Even if it be an accident, it must have a cause beyond the realm of energy. The source that generated motion in the inertial energy must overlie beyond energy. And it must be dynamic as well. Also, as energy does not decay and is not subject to any loss or gain, the said source which effected motion must enjoy an independent existence of its own.

We also observe a superbly organised order in universe. Such a well organised order betrays high intelligence and wisdom in its planning and execution. This speciality is manifest from the micro to macro level. The universe is ticking like an atomic clock perfect to the fraction of a second. It maintains a mathematical accuracy. Its accuracy guarantees the presence of a particular star or planet to be at a specific location in space at a specific time. Assured of such an organisation and its beat the astronomers can calculate and predict many celestial events well in advance and charter the space traveller's route and destinations. Such a built-in system betrays wisdom and cannot be a part or attribute of the inertial energy. Consequently the same cannot manifest from any process exclusively limited to or simply a transformation of energy. For such a manifestation there has to be a source independent of its material source which energy is.

Another notable character of universe is that it grows as one unified system. Everywhere we find abundance. Whatever gets produced is manifest abundantly. And yet there is no waste production. The universe is a unified system rich in abundance. Each branch is a part of a bigger unit which in turn is unified with a still larger group or family and so on. Each unit from the lowest or the smallest to the highest or the biggest is integrated with the other as well as with the whole. Even the opposites form part of the same unified system. Such unity in multiple abundance forming one whole system can hardly be explained to have resulted out of energy on its own.

We know that all universe is matter manifested and organised in multiple varieties and their combinations. The energy has become transformed in all the bodies of matter. But we also know that the properties of its product matter widely vary, and such a variation makes one product characteristically different from the other. This has caused wide dissimilarity, and even similars are not same in composition.

Modern science treats them as chemical reactions or interactions as the case may be. But how is it that even though such chemical reactions and interactions inbuilt within the product matter differ widely yet they cover the multiple facets of contributions towards building up a cosmic whole system, called universe. Is it that the energy having become kinetic through some so far unknown cause to modern science has run into an even doubtful one billionth chance of converging upon a superbly organised whole system of creation, or is it that the unintelligent inertial energy flies through a pattern or action plan devised by a supreme source of wisdom, knowledge and dynamism, dancing to its tune as propelled by the same? It is incumbent upon the world of modern science to take a more pragmatic approach and view the mystery of creation from the Vedic angle.

Summing up this discussion it may be concluded that the following characteristics manifest in universe are not explained by mere acceptance of inertial energy as the fundamental cause of creation: appearance of motion, superb orderly organisation, its growth as one integrated system, and its wholesomeness including a wide complexity of intelligent features. The fundamental law is that something cannot emerge out of nothing. The characteristics manifest in universe and enumerated above are hard put to be determined through their origin from the said energy. They all assert in favour of a dynamic source full of wisdom and knowledge being in existence beyond the energy to which all matter owes its creation. Such a source will, *ipso facto*, have to be infinite, over-reaching and exceeding the limits of energy.

The state as described above may well be finalised in naming the attributes of such a perception to (1) existence, (2) wisdom/intelligence, (3) dynamism and (4) a completely wholesome state endowed with perfect efficiency in all scienticfic knowledge including mathematics. Veda takes predominant cognizance of such an extra-ordinary being. As we shall see from a quote, the Yoga treatise literally calls it an extraordinary being. But first let us end this analysis by reminding that the Vedic term for eternal existence is sat, intelligence and dynamism both are fully covered by cit and the whole as described at (4) above by the single word ānanda. And the three join together making it *sat+cit+ānanda* = *saccidānanda* to give such a being a name based upon Its attributes. It has no mass, no form, hence no limitations of any kind whatsoever. Let us try to have a glimpse of this greatest scientific truth stated by Veda through an understanding of the three Sanskrit terms one by one below.

Sat: Literally it denotes eternal presence/existence. It is unborn. It always, eternally, within and without time, is or exists. Nothing exists without It save the energy. It enjoys supreme existence, i.e., Its existence surpasses each and every other existence which remains subordinate to the former. It is beyond any measure. Its extent and proportion is simply measureless. The biggest measure that science has come to know is universe which hardly covers a part of that existence. It does not move, as there is no room left out for It even to stir not to say of moving. Being present everywhere It stands witness to whatever else exists or occurs. It exists even beyond the scope of mathematical calculations. You may name It infinite through a symbol of infinity. But then, how about a point? A point has no size. It is simply a location. But a location has existence, and as existence It is present there Itself. Hence mathematics may at best describe It as an existence from point to infinity. This will become a chain of continuous rings or circles beginning with the location point centre extending upto merging with the one conceived at infinity. And yet the concept of existence may extend still further surpassing such infinities till you are left with no alterative but to concede it as one encompassing and surpassing all. But even here you are expressing within the limitations of both the language as well as the mind. That is why the *ṛṣi* says that Its existence may at best be realised but can never be fully known. '*Neti, neti*' : no end, no end. The limitations of the mind and the creation are far too small to cover the extent of Its existence. But it may be realised that It is and exists as such with no limiting circumference.

Cit: It denotes consciousness or the quality of being awake, i.e., fully alive both intellectually as well as for any action. The word *cit* is derived from root *citī* which denotes consciousness as above. Our dictionary confirms the meaning. This term, therefore, covers what we understand at physical level as mental prowess as well as capability to act. But here the first state *sat* surpasses the limitations of both physical and mental levels. Hence the *cit* state denotes wisdom and dynamism in totality. In other word,s no knowledge and no action is beyond it. It also covers the ability to lay down any order and to cause some one, say the inertial Prakṛti, to arrange its own self in that order and perform accordingly. In other words, It lays down the constitution of the cosmos, the rules of its maintenance and working, called *dharma*, including the entire process of creation. It directs *Prakṛti* and causes the same to perform to the plan set by It.

Ānanda: Derived from the root *nand* denoting *samṛddhi* meaning great prosperity, growth, thriving, welfare, fortune, perfection, excellence, abundance, wealth, riches, etc., with the prefix *ā* denoting complete inclusion of limitations all around and the suffix *a* to form a noun, the term includes to mean all of these and much more. In short, it conveys the sense of prosperity and abundance as well as excellence to the limit supreme which transcends all other limits.

All prosperity is inbuilt in matter. It is also matter which imparts joy and pleasure to the biological life. It is, therefore, the reaction related to its manifestation and the reaction related to its multiple combinations which is inherent in matter enabling it to cause such effects. The inertia cannot cause to produce such joyful, pleasant or otherwise effects. It cannot on its own display pattern of interaction which may contribute to the growth of a system that is a whole in itself and kicking.

In short, *Saccidānanda* denotes the Being who exists everywhere, enjoys an all-pervading all-encompassing and all-surpassing existence, is all dynamic, supremely intelligent and wise, is a self contained whole containing all knowledge and is inclusive of all limits, of all sciences and skills. It has no body, figure or form because these all suggest limitations. Such is the name given by Veda to the Extraordinary Being of Yoga. It has no name of its own. It is known by nouns based upon its numerous attributes. It is called *nirguṇa* zero attribute. It is noteworthy that a zero in Vedic concept represents a whole. In R.V. 10/129/3, it has been called *tuccha* which is a synonym of *śūnya*, i.e., zero. It is the Supreme Master, that is why it is called *Īśa* or *Īśvara*. Even the noun *God* in English language is a corrupted form of a Sanskrit formation *goḍa*, derived from the root *guḍ* to guard or protect, which would mean the guardian or the protector of universe.

There are two fundamental verbs in the human language: *to be* and *to do*. They cover the entire linguistic spectrum of existence and activity. It is to be all and do all as every kind of activity finds its final source both in appearance and action in It. It causes all actions and yet It remains a non-doer. As the *Puruṣa* described in an earlier chapter, It simply leads or directs *Prakṛti*, energy, onto the process of creation, maintenance and or dissolution as the case may be. It imparts a purpose to *Prakṛti* and stimulates the latter to perform, to achieve the same. Naturally *Prakṛti* remains the principal doer in physical sense. But It (the *Saccidānanda*) simply remains the efficient cause and a *sākṣī* or witness to the activity of the energised *Prakṛti*.

Let us consider the two references describing the extraordinary being indicated above. The following is a *mantra* from Yajurveda:

sa paryagāt chukramakāyamavraṇamasnāviraṁ śuddhamapāpaviddham
kavirmanīṣī paribhūḥ svayambhūr yāthātathyato'rthān vyadadhācchāśvatībhyaḥ samābhyaḥ

(Y.V. 40/8)

(sa) That Supreme Being **(pari+agāt)** which pervades and surrounds the entire universe, **(śukram)** is bright, resplendent, spotless, **(akāyam)** incorporeal, **(avraṇam)** unscarred, **(asnāviram)** without sinews, **(śuddham)** pure, faultless, exact, according to rule, unqualified, complete, accurate, **(apāpaviddham)** unpierced by any evil, **(kaviḥ)** possesses insight, is enlightened, **(manīṣī)** learned, intelligent, **(paribhūḥ)** is all around, covers every bit, **(svayambhūr)** self-existent, **(yāthātathyataḥ arthān vyadadhāt)** has imbued all objects with properties, actions/interactions and nature in accordance with their creations **(śāśvatībhyaḥ samābhyaḥ)** for eternal years.

The second reference is as follows:

kleśakarmavipākāśayairaparāmṛṣṭaḥ
puruṣaviśeṣaḥ īśvaraḥ

(Pat. Yog. 1/24)

Untouched by **(kleśa)** afflictions, distress, anguish, **(karma)** actions, **(vipāka)** effects, results, consequences thereof **(āśaya)**, desire

(aparāmṛṣṭaḥ) (puruṣaviśeṣaḥ) an extraordinary Puruṣa or being **(īśvaraḥ)** is *īśvara*, the Supreme Lord.

Let us add a third reference from Śve Upaniṣad:

na tasya kāryam karaṇam ca vidyate na tatsamaścāpyadhikaśca dṛśyate
parāsya śaktirvividhaiva śrūyate svābhāvikī jñānabalakriyā ca.
(Śve. U. 6/8)

(na tasya kāryam) It has neither any work to perform **(karaṇam ca vidyate)** nor there are in existence Its implements **(na tatsamaśca)** nor there is any one Its equal **(apyadhikaśca dṛśyate)** nor is there visible or known to be anyone exceeding It. **(parāsya śaktir vividhaiva śrūyate)** The excellence of Its power is heard or known from various sources **(svābhāvikī jñānabalakriyā ca)** and Its knowledge, force and dynamism are natural states of Its being.

This is a brief outline how Veda presents the state of the efficient cause as a subject of science. Each of the four Vedas presents this Supreme Being by more than one name. Such names are taken to denote various *devas* or so-called gods by those who are keen to establish their earthly wellbeing and prosperity through the practice of idol worship. But each Veda at some point in its text forcefully states that all such names denote one and the only Supreme Being lest there be any confusion on the issue. *Ṛgveda* specifies - **ekaṁ sad viprāḥ bahudhā vadanti** (R.V. 1/164/46). The eternal is only one but is described by the men of wisdom in many ways. *Yajurveda* says: **na tasya pratimāsti** (Y.V.32/3). It has no idol or duplicate.

Another angle is the gender of the Supreme Being. Probably the Vedic text and the Vedic literature is the sole exception in human expression which depicts a scientific outlook in this matter. Normally God is described in masculine terms. There is no rhyme or reason for it, but it is customary, probably because of the male dominated society. But Veda speaks of that Supreme Being in all genders, i.e., masculine, feminine and neuter. Quite a few Vedic nouns related to It, are in masculine form, others in feminine, and still others in neuter form under the rules of grammar. Generally the objects betraying intelligence and emotions are treated as masculine or feminine and the non–live ones as neuter, although there may be any number of exceptions. In a language like Hindi in India, the gender is expressed by the verb, and there is no neuter gender. But the source of intelligence and emotions in the creation could hardly be the sole basis for manifestation of inertial energy, and appears to go beyond the same, so that the same might have been qualified as masculine or feminine. On the other hand, all unintelligent and unemotional objects being transformations out of inert origin, although under divine directions, justify the neuter gender qualification. But this text has preferred the use of neuter singular third person pronoun, i.e., It for the same.

Another Vedic name for *It* is *Om*. A combination of two vowels and one consonant a+u+m=Om still retains its hallowed position in human territories upon this globe even beyond the realm of Veda. The three letters that combine its constitution denote the three attributes of *Sachidānanda*, i.e., *sat*, *cit* and *ānanda* respectively on the basis of alphabetical analysis. Moreover its etymological interpretation claims its derivation from the root *ava* denoting 19 meanings enumerated specifically by Pāṇini, which in combination with prefixes would cover a field of hundreds of meanings, probably the greatest field covered by any single root. It also enjoys a lone exception both in the Vedic as well as the popular form of the Sanskrit language. *Om* is exempted from subordination to all rules of grammar. This is to say that under any kind of usage its form would not be modified by any of the prefixes or suffixes and would remain unchanged or unaltered under any context or reference. This is the manner in which the most systematized language, namely, Sanskrit and Veda pay their homage to the Absolute Truth and Supreme Master of Sciences. *Om*, in a slightly changed form of *Āmīna* or *Omen*, is still in vogue at the end of prayers and solemn affirmations both in Christian and Islamic worlds.

It is time now for the modern science to realise the overwhelming supreme position of the efficient cause which activates the material cause, i.e., the energy to produce this universe as a self-contained whole system. The inertial energy, by itself, under the rules of science and its very nature as well, cannot deliver the goods. The incorporeal Supreme Master of sciences, wisdom and dynamism, by Its very nature, cannot bypass, overshoot or make any exception to the perimeter of the scientific approach. It is certainly not, nor can it ever fill the image of a traditional or conventional God who incarnates Himself or sends His son or Messenger to help mankind expiate their sins, destroy the unbelievers or reward the believers and devotees. The universe behaves under the rules and system of science; its material cause, *Prakṛti* or energy and its Supreme efficient cause both are bound by their very nature. The Supreme Being, named *Saccidānanda* by Veda cannot act otherwise. As Maharṣi Dayānanda put it, the Supreme Being cannot create another Supreme Being or Its Duplicate, nor can It favour some and frown upon others. The system created by It is called *dharma.* This *dharma* takes care of the creation. This chapter ends with a Sanskrit quote: *dharma eva hato hanti, dharmo rakṣati rakṣitaḥ.* It says: The *dharma*, i.e., the system of creation and its rules, if violated, recoil upon the violators; it preserves those who protect them.

— 19 —

KĀLA

In Vedic thought the concept of *deśakāla* is universal. It is a compound of two nouns. The first one, i.e., *deśa* denotes a field or territory which is three-dimensional in mathematical terms. Coupled with the same the second one kāla (time) gives us the Einsteinian formula. The fact is that the concept of calculating time as a measure did not exist in pre-Einstein period of modern science. It was only after that renowned scientist had accomplished the extension of the field of physics by installing the study of energy at its supermost position, that due to him the concept of time also underwent a fundamental change. But in India the emergence of science occurred from Veda and Veda not only places the advent of *ṛta*, K.E., ahead of *satya*, matter, but even bases the definition of *satya* on being unexceptionally genuine in *sat* which *ṛta* is as Prakṛti iṇ motion. In science based upon Veda the fundamental subject of study is *ṛta*; *satya*, as matter, enjoys a secondary position and that too, in the context of *ṛta*. Thus the basis of the scientific thinking and calculation from the very first day has continued to be *deśa+kāla*. It is a different matter that, in the course of time, having gone astray from its Vedic path, the scientific base went on shrinking. But even then, maybe simply on the basis of tradition, this term *deśakāla* continues to survive in social behaviour.

Under the British rule, the teachers in India used to acquaint their students with four directions. We are a quarter short in the sixth decade of our independence, but the lesson still continues to be taught the same way. Only the students of science and mathematics are made conscious of three-dimensional measurements. But Veda addresses even its prayer in six directions which clearly establishes the three-dimensional base to the thought process. The calculations of the celestial journeys of the planets and such other bodies along their respective orbital courses formed the body matter of old calendars. The Vedic scholars were fully aware of the combined mathematics of both *deśa* as well as kāla, and such a knowledge was not confined to Veda itself, but was the main stay of their entire astronomical calculations and researches. They were not only fully acquainted with *kāla*, time, but were no less wise regarding *a+kāla*, timeless state of existence.

The noun *kāla* is derived from root *kal* and is an abstract noun obtained with the addition of suffix *gha*. The root *kal* has been mentioned four times separately out of which two are irrelevant here. But the first one specified to denote *śabdasaṅkhyānayoḥ* and the last as, *gatau saṅkhyāne ca* are pertinent. In both the denotations *saṅkhyāna* is common. The meaning denoted by *śabda* in the former and *gati* in the latter are purposeful. The noun *śabda* denotes manifestation and gati communicates knowledge, motion and accomplishment. But saṅkhyāna is simply counting or measurement. The derivation of *kāla* from the first root will denote manifestation as also the period consumed therein, while that from the second root will specify distance or amount of motion covered or may be covered in its efficient accomplishment. The concept of the measure of time is inbuilt in the very root. Moreover this position is unambiguously maintained throughout in Vedic literature. If the creation of universe, its maintenance and dissolution is seen as an eternal flow, all of it, is occurring as a current account in the flow of time.

Let us also consider a bit more about the implication of the word *saṅkhyāna* (counting). Each and every event occurs in time, and there always happens to be some interval between the preceding and the following event, be it howsoever insignificant or of short duration. In the process or flow of creation an interval occurring between any two events is a must. The direction of creative process is always forward, i.e., in the direction of motion. Hence if the interval does not appear in a position fore or aft, it shall be on the right or left or up and down. In a three-dimensional field every direction stretches forward. And in each direction one event will thus be related to the other. This relativity occurs and is maintained in all of the six directions along the basis of the three axes which run perpendicular to each other at each point. Consequently for the entire series of the events to occur, the universe has to have a field of occurrence. Such a field has been named *ākāśa* by Veda. Hence *ākāśa* is not merely the field for the activity of creation, it is also the field of time. In fact the knowledge of events and their intervals is the knowledge of time. That is why time has been associated with *Saṅkhyāna* (counting). And an interval is the difference/distance occurring between the two events. They, therefore, form the two sides of the same coin. This peculiarity when dawned upon Albert Einstein, he introduced time to Modern Physics as the fourth dimension. To the Vedic knowledge it is an integral part.

The creation is a totality of the series of events occurring in *ākāśa*. It is carried out by *ṛta* which, through its motion, transforming or making itself more compact, finally accomplishing the image of matter, proceeds on its motion. In this three-dimensional field of *ākāśa* the perpetuity of motion initiated with the rise of ṛta occurs like a dotted line of events stretching in a direction and continues that way. Such an imaginary line of dot-like events marks time. The lines indicating the three-dimensions are called axes. Modern science calls them x, y, and z axes. The x denotes east-west, the y north-south, and the z up and down. The time is denoted by letter t. The point at which the three dimensional lines meet each other at right angles, is the centre point of the three dimensions as well as that of time. A circle may be drawn taking any point as the centre, wherefrom each point on the circumference will be equi-distant, but in the field named *ākāśa*, there are three-dimensional centre points everywhere, but no circumferences. It is adjustable alongwith the expanse of *Prakṛti*, energy, which has no circumference. All limitations are confined within the expanse of *ākāśa*.

It has been stated on more than one occasion that according to both Veda as well as modern science, creation is a result of energy in motion. The entire universe is created by the kinetic energy. No two events occur simultaneously at the same point of occurrence. If they happen simultaneously, they must have different centres. On the contrary, if they appear to have the same centre, the two must have a difference of order. The occurrence of events being under a constant motion, sameness of centre is merely an illusion, not reality, because the centres of occurrence keep on changing constantly. Thus the entire process of creation remains effective on the plane of relativity one way or the other.

Hence, to know any situation or position in respect of the creation of universe, it is incumbent to locate or determine the point of its occurrence or happening. This is possible through the knowledge of motion under which the position or the situation desired did occur or is going to occur in this vast expanse of *ākāśa*. The subject of such a knowledge about motion is time. On the basis of the passage of time modern astronomers or those of the Vedic age could gain the knowledge of stars, star-clusters, planets or any other body or group of bodies in ākāśa or regarding any event that happened or was going to happen to them in the past, present or future calculated from any point of time. Upon our earth the most common example has been the timings of solar and lunar eclipses. From that point the modern science has made great strides. It is successfully employing the knowledge of motion gained to determine a route to the desired planet or body, to approach and photograph the same, and even to land a spacecraft upon the same. Of course the knowledge of motion includes all the factors or forces that affect the motion and the ability to work out the differences through proper calculations.

Innumerable motions are on in the world of creation. It has been mentioned earlier that billions of galaxies are on the move, each of which contains millions or billions of stars. So many of them are like our sun, and quite a few far bigger than it. All of them are racing along their respective tracks in their respective galaxies. Those of them which have their planetary systems with them are dragging the same along with each planet orbiting its own star-sun. Then there are comets which keep on appearing at regular intervals, and going round the blazing star again disappearing in the depths of *ākāśa*. The planets may have their satellite moons. In addition to all of these the probability of planets like our earth existing elsewhere is neither denied by Veda nor by modern science now. On the contrary the Vedic note is quite positive in this respect, that life is not neccessarily limited to earth and our home planet is no exception. On account of life innumerable kinds of motions keep on occurring upon this earth alone, in addition to the motions of this carrier planet of ours. Even the imagination shudders when the probability is considered. Then we know there are other bodies, like meteoroids, rushing away. Some basis is essential to measure and account for all sorts of motions.

Our earth is orbiting our sun. All other planets are doing the same. The completion of one orbit around the sun is called one *saṁvatsara* by Veda. In other words an orbit is a *saṁvatsara* cycle. All celestial bodies have their respective *saṁvatsaras*, the planets around the sun and the stars around their galaxies. A *saṁvatsara* is the name of time taken in completing an orbital round. It differs from planet to planet and star to star. The *saṁvatsara* of our earth is counted one year in time. Such *saṁvatsara* of each body, be it a comet, a planet or star/sun is the completion time of one round done by it. Like the earth, other planets of our solar system are orbiting the sun. They too, have their sunrises and sunsets, as well as days and nights. Their velocities being known or having been calculated, the length of days and nights occurring there, can be determined. They may further be subdivided into shorter units of time.

The ancient astronomers noted that the rising of stars in east and their setting in west appeared so due to earth's rotation upon its axis. They also observed that a few of the stars in the north nearer to the horizon, appeared to travel from west to east. The Vedic *ṛṣis* were fully aware of earth's rotation. Not only that they were aware of the phenomenon that each celestial body, while moving on its course, keeps on rotating upon its axis. The position of *dhruva* (pole star) is indicated by its nomenclature as the noun *dhruva* means steadfast. Through a keen observation with their bare eyes they could determine that the axis of our earth orbits the polar star at a distance of one degree.

The Vedic seers based their calculation of time upon the principle that if the velocity of a body is known, the time it would take to cover a definite distance, and if the time is known then the distance covered can be calculated. Let us take an example from our life. A train is crossing a bridge. If the velocity of the train is known as well as the length of the bridge then considering the length of the train the time of its crossing the bridge can be determined. The calculation may as well be done in reverse order by treating the train standing still and the bridge mobile. Such a determination is called *relative velocity* and the answer is always the same. Vedic astronomers adopted this mode of reverse calculation by treating this earth as still on account of their own position. It was easily workable and practical. In course of time when others learnt the technique of calculation, they came to believe the earth to be stationary and the rest in motion. In the same form the method got included in the Testament and the Church treating the same as divine knowledge bestowed sanctity upon it and taught accordingly. That is why when Galileo, in defiance of that religious belief, declared the earth to be going round the sun, he had to face the ire of the Church. Even today, in spite of numerous modern equipments the *modus operandi* in practice is the same ancient Vedic age one. In metaphorical language the train is treated as stationary and the calculation is done taking the bridge to be moving.

Motion is common to all the bodies in universe; the difference lies in their velocities. It is, therefore, neccessary to understand the process of motion. For this purpose let us take our own earth. Our moon is moving completing its orbit around the earth in about 29.5 earth days. The earth is going round our sun with this lord of the night in tow, and the sun is on the move on its track in the Milky Way galaxy. The result is that along with our sun all the planets of this solar system tied to it including our earth, together with their respective satellite moons, are in forward motion as well. Due to this arrangement of motion, the moon certainly continues to orbit the earth, but it never returns to that momentary location point which it passed 29.5 earth days earlier. The real position is that at every moment of its onward movement along its orbital path, it is, of course completing its cycle from the point it left approximately 29.5 days prior, but instead of returning to that specific earlier momentary location point, it arrives at a point in its path which is, in alignment with the earth's motion, along the latter's orbital path around the sun. The same rule holds good in respect of all other moons of the respective planets. All of them are completing their respective orbits without returning to any of the points on their passage, but continuing their run along the coil-like orbital path respectively in line with the sun. The same system is effective with all of the stars/suns including our own which is tied to the border of our galaxy. They are running in their respective orbits adjusting their tracks every moment accordingly along new location points. In short, not a single body in the entire universe, though constantly on the move, ever returns to any of the location points of its previous round. Each orbital path is shaped like an invisible spring of which every spiral in spite of being identical to it, does not cut the preceding one but runs along a point ahead or above. All over and everywhere in universe this orbital motion is occurring along the same spiraling movement. Because of this peculiarity none of the bodies returns to any location point, once has crossed. It always treads upon a new path.

By taking a fix of star positions, Vedic astronomers knew that a star or a constellation, when a fix of its position is taken again at any particular point, arrives on that point the following night slightly later than the moment of the previous night. Keeping such a difference in mind they worked out what is now called the sidereal time or the star time. Accordingly the sidereal day, month and year were determined. Such a calculation is done on the basis of *saṁvatsara* period and the position of the sun in relation to certain stars. The method being adopted by the modern astronomers today is no different. The difference between the timings of the arrival of a star at a fixed point on two successive nights approximates four mintues. The sidereal time is calculated today keeping that difference in mind.

When Columbus sailed westwards, the Europeans believed this earth to be flat, and some of them were even afraid that his ships and the crew might not be lost down some enormous sea-fall at the end of the earth. But Veda since the very first day has called the earth *bhūgola*, born like a ball. There was no misgiving about the shape of our home planet. Not only that, Veda also states the noun *khagola* (ball-like heavens). Like the division of this globe with imaginary latitude and longitude lines, and determination of equator and tropics of cancer and capricorn the same was done with *khagola* celestial globe. Like the earth, the celestial poles were fixed.

Simultaneously the time was divided upto so small a unit which comes to smaller than a millionth part of a second. The whole system of counting was established. Modern science has adopted the same. If there is any difference it is merely in changes introduced by science in smaller divisions of time. But even in such divisions the basis of time division given by Veda has come to be adopted. This fact is established by two different manners the small unit of time is pronounced in English language. The word *minute* is spoken in two different ways to convey two different meanings. It is pronounced *min / it to* denote sixtieth part of an hour, and it is pronounced *mī nōōt / mī nyōōt* meaning extremely small in size. At the time of such a determination at a papal congregation even a sixtieth part of an hour was considered a very small time. But when later on need was felt to have a still smaller division of time, the *minute* was again divided into sixty parts and a single part was named *second*. The very nomenclature *second* discloses the history of being so named as it was determined as second *mī nōōt / mī nyōōt* unit of time.

Moreover division of time into sixty parts is a pure Indian mode wherefrom it has been adopted by the West. Such a division of time to indicate finer units is seen in the determination of *pala*, *vipala* which can be traced to a fairly ancient period. In India, even a daynight period has been divided into sixty parts, each unit being called a *ghaṭī*. The west has divided the same into 24 parts, each an hour. The division of a circle into 360 parts is a direct Vedic proposition. The entire global and astronomical calculation of modern science are based upon the same. And why is a circle divided into 360 parts or degrees? Because the earth's daily angular velocity unit is one degree and it completes the round in 360 units, each of which is considered as a daynight unit (not rotational).

In the universe each body is in motion under the influence of centripetal and centrifugal forces. On account of the two forces a body moves along an elliptical path, which it follows in a spiraling manner. Hence a value of motion came to be determined in tune with the elliptical *saṁvatsara* divided into 360 parts. This is a system given by Veda and the same has been adopted as it is by modern science. The other value was established in terms of distance for which the addition or multiplication of the unit distance was understood and adjusted with the language of time. Veda determines the distance in the language of time. As the entire series of events related to creation occurs in motion, it is easy to calculate the time taken by the happening or its age in terms of time which in fact amounts to distance covered at a specific velocity.

In India when the concept of time was worked out on a fine scientific basis it was considered from a particle to the entire duration of the cosmos, from creation to dissolution. Reference is found in *Śrīmadbhāgavata* at 3/11/4 which says: the *kāla*, time, is called *paramāṇu* which enjoys particle state, while that which enjoys all the states from creation to dissolusion is said to be absolutely mega time.

It has already been stated that the entire *Brahmāṇḍa* is in motion. From micro to macro, every bit of it, is mobile. No part of it is stationary. Motion is relative and varies from the extremely slow to amazingly fast. And time is the sole measure of all sorts of motions. It is not easy to fully understand the entire gamut of motion and its play through any other medium than time.

The *saṁvatsara* already discussed earlier is the nature's clock of time and distance both. It emerged with the birth of the very first atom, an electron doing rounds of a proton nucleus as the atomic *saṁvatsara*. Veda has stated:

samudrādarṇavādadhisaṁvatsaro ajāyata (R.V. 10/190/2)

It means: from the vast ocean of radiation and matter, and based upon the same, appeared *saṁvatsara*. Such appearance was based upon creation of Hydrogen matter in which even an electron orbiting a proton nucleus was doing rounds under the grip of the dual forces, centripetal and centrifugal. The *saṁvatsara* phenomenon is universal. The planets of each and every star-sun are orbiting the latter. The time each planet takes in completing its orbit is the period of its own *saṁvatsara*. The sight is familiar in our solar system. Our earth is performing the same in its own *saṁvatsara* time. Its period is definite and it does not vary in routine. Hence it (*saṁvatsara*) is the natural clock which indicates time in the universe and is associated with and governs practically every motion under the process of creation and maintenance. The Vedic wisdom has based its calculation of time upon this *saṁvatsara* since the beginning.

The Vedic ṛṣis, seers, determined the base unit of time to be a single combination of day and night and classified it in four categories. The measure of a *saṁvatsara* was taken at the completion of 360^0 on the orbital path and one degree was a single day and night whole unit. The four classifications were the human year and its day-night units, the lunar year's day-night units, the *deva* year with its day-night units and finally the *Brāhma* year and its day-night units. All the four were calculated on the basis of respective *saṁvatsaras*.

For practical reasons two exceptions were also provided. As the rotational day-night occurrence differed from the 360^0 count of the *saṁvatsara*, the sub-divisions of time determined were also slightly adjusted to the rotational day-night period, i.e., 24 hours or 30 *muhūrtas*. Similarly as the lunar year did not confirm the earth *saṁvatsara* an additional thirteenth month was provided to be observed each 2.5 years in a manner to balance the two. But the new year began with the sun rising on the vernal equinox, a sure proof of base determination, The third *deva* day-night unit is the six month day and night alternate in the arctic and antarctic regions. The two are called *uttarāyaṇa*, sun in the north, and *dakṣiṇāyana*, sun in the south. *Deva* denotes the sun, so it is a solar day-night period which is equal to one human year. Naturally a solar year equals a period of 360 human years *saṁvatsara* wise.

The fourth and the last classification is related with universe. That is why it is named after *Brahmā*, the atom. It runs through the universal calendar. It begins with the advent of *Brahmā* and carries the count of time till its dissolution. The period of one earth *saṁvatsara* for such a huge calculation was rather insignificant. So the point of time of the emergence of the system of planets was taken into consideration for calculating the *saṁvatsara* being done by the mass of matter that went to form various planets and in that form continues to orbit the sun. The Veda speaks about a *yuti* (conjunction) of the planets of our solar system occurring at an interval of 1200 solar years' periodicity which is equivalent to a duration of 432,000 human years. This *yuti* period from one planetary conjunction to the other has been declared as a single unit of measure in time and has been named *kaliyuga*.

The author of the famous book *Vedic Sampatti*, i.e., *Vedic Wealth*, has discussed the subject of this *yuti* periodicity in some detail. He has also quoted from another book which we reproduce below:

> "According to the astronomical calculation of the Hindus, the present period of the world, Kaliyuga, commenced 3,102 years before the birth of Christ on the 20th February at 2 hours, 27 minutes and 30 seconds, the time being thus calculated to minutes and seconds. They say that a conjunction of planets then took place. The calculation of the Brahmins is so exactly confirmed by our own astronomical tables that nothing but actual observation could have given so correspondent a result."
>
> - (*Theogony of Hindus*, by Count Bjornstjerns, p. 32)

This quotation speaks of a confirmation of the Vedic position regarding the commencement of the present period of the prevalent *kaliyuga* from the occurrence of a '*yuti*', i.e., planetary conjunction we are mentioning.

This unit period of 1200 solar years or 432,000 human years is called *kaliyuga*. The word *kali*, according to our dictionary is a symbolic expression for number 1, and *yuga* is stated to mean long mundane period of years. This is the unit adopted by Veda to measure long periods in the life of our earth as well as that of universe. As *kaliyuga* denotes one unit, two *kaliyugas* are named *dvāpara*, three *tretā* and four *kṛta*. The word *yuga* may or may not be added. In all these four, units are determined, which are collectively called a *caturyugī*, a unit of four *yugas* which are equal to a total period of 10 *kaliyugas* or 12000 solar years, i.e., 432,0000 human years. The *day of Brahmā* which is also called a *kalpa* is of 1000 *caturyugīs*. Equal is its night, also named a *kalpa*. A period of 360 double *kalpas* forms one year and 100 such years the life or existence of *Brahmā*. According to Veda the age of universe, works out in human years to $864 \times 36 \times 10^{10}$. It is specifically pointed out that the period of a *kalpa* is mistakenly taken to be the period of creation. It is simply equivalent to twelve hours period of the universe in terms of human span of life of 100 years. It may further be added that *Brahmā* has completed 50 years of its life. The present time is its first day of the 51st year.

A mistake has come to stay in the calculation base of a '*kalpa*' on account of treating it as a matter of faith rather than mathematics. The period of *kapla* is a *day of Brāhmā*. Like the human, lunar, and solar days the *Brāmha day* ought also to have been divided in 15 parts. In a human day such a 15th sub-division is called a *mūhurta* and in a lunar day it is known as a *tithi*. In respect of the *Brāhma* day such a sub-division was, probably, named a *manvantara*. This is not a Vedic sub-division. In the manuscripts only 14 names of the '*manvantaras*' are available. Veda has specified the duration of a *Brāhma day* as 1000 *caturyugīs* which come to 12000x1000= $12x10^6$ solar years or $432x10^7$ human years. Neither figure is divisible by 14 even upto decimal places. On the contrary both are divisible by 15. But instead of treating it a human error, most probably in copying a manuscript or by accident and additional manuscripts being not available for cross-checking due to some mishap, all sorts of illogical explanations have been made out to justify what is manifestly wrong. What should have been done was to search out the 15th name or call it an unnamed *manvantara* and clarify the mistake? So far as Vedas are concerned the period declared by them stands firm.

According to the modern science the present creation started some 16 billion years ago, and it is still in its youth. The time calculation as per Veda mentioned above takes us to approximately 2 billion years in the past for present *kalpa* to commence. Another 2 billion plus years still remain in balance to follow. In both respective positions this much does appear common that the creation is passing through its youth, but the difference between two estimates regarding the birth of Creation is enormous. It is noteworthy that until the 19th century of Christian era, the West was unwilling to accept Vedic figures as they were considered too high. And today the same science having developed further, disbelieves the stand on the ground that Vedic statement falls far short of their own estimates now. Let us review this position.

The science books published in 1950 had brought the estimate of the birth of universe to about 2 billion years, and their writers expressed surprise as to how the Indian scholars during the ancient Vedic age could establish such an age to be nearly 2 billion years. Today the same science is taking its estimate to 16 billion years. Well, has the science finalised its calculation? Science does not admit it. According to it, there is every possibility of the period of the origin of Creation going up, and it does not overrule inordinate increase. Our submission is that Vedas are quite ancient, a statement accepted by all. How much ancient they are, is not under consideration here? But it is a fact that there is a world of difference about the antiquity of Veda between Indian and Western scholars. Be it as it is but even if the Western estimates about the antiquity of Veda is taken for the sake of argument then at that time how could even a declaration that the present kalpa commenced 2 billion years ago be imaginary? In that age even mathematics was not developed enough in the West to be able to count that number. Only 1400 years ago in the land of Arabia, at the time of the birth of Islam, it was declared that the world would last 14 centuries more, after which there would be the final end. Those 14 centuries have elapsed, but there is no trace of the end. The European scholars themselves had been sweating in determining any period prior to Christ.

The fact that must be taken into consideration is that the period of the origin of Creation given by Veda is based upon mathematical calculation, and such a huge number could only be arrived at in that age on the basis of some calculation. A conclusion reached through mathematics is always scientific and in no case the result of any wild flight of imagination. That is why we have been at pains to describe and narrate the whole order of time behind such conclusions. Even if for the sake of argument it is accepted that the modern scientific assessments are right, there is no escape from the conclusion that Vedic seers, too, had arrived at their period of *kalpa* or the origin of Creation through calculations conducted along a scientific basis. In every scientific line of thought it is its scientific basis which is open for as many re-checks as desired. Same is the position with the modern science. That is why it cannot be blamed for its earlier declarations and conclusions. On the contrary it deserves accolade because it has been retesting its scientific base all along. Then, why should Veda be not extended the same courtesy? Why should any religious bias or even any so called scientific finding be allowed to override our scientific temperament and approach? What is of paramount importance is to provide a scientific basis to our thinking and researches related to creation, and re-evaluation of our process along the same basis.

If the period of the origin of Creation as stated by Veda is considered in the light and spirit discussed here, all of the periodic assessments mentioned above come under the ambit of deserving reconsideration. No scientific thought will conceive a human figure named *Brahmā* for conducting the process of creation. That *Brahmā* denotes atom has already been clarified earlier (chapter 6). Hence the birth of *Brahmā* means the birth of Hydrogen atom. What is needed is application of scientific analysis.

It is noteworthy that Veda declares an eternal cycle of creation and dissolution. It may also be noted that *pralaya*, dissolution, has been stated following each *kalpa*, a day of Brahmā, but the *mahāpralaya*, the great dissolution, occurs only after the completion of 100 years by it. This simply denotes the state of the final dissolution of all matter into energy leaving hardly any balance in atomic form, a state that comes after the 100 years of *Brahmā* are completed. The inference is that in dissolutions stated during its life time the atomic form keeps surviving. And atomic survival means the matter as we find in the universe. This aspect will be further discussed in a full chapter on *Pralaya* which follows. *Brahmā* has completed 50 years which approximates 16×10^{12} human years. Thus, if the age of the universe is to be calculated on the basis of Hydrogen atom, Veda still provides enormous leeway to Modern Physics for adjustment and correction.

Some people point out that the Vedic figures appear to be conveying exact numbers while modern science, in spite of enormous increase in its assessments, speaks in approximations, e.g., according to Veda a *kalpa* lasts 4.32 billion years out of which over 1.96+ billion years are gone, while science assesses the universe to be about 16 billion years old. How can Veda be so specific? The reason is that all of the Vedic figures are based upon astronomical calculations, more particularly the *saṁvatsara*. The astronomical calculations have to be exact in time even beyond a second. Thus all Vedic figures of various periods or durations are the multiple of *saṁvatsara* period which does not permit any approximation unless you do it deliberately. Modern science has developed far superior instruments of observation than Vedic scholars.

The irony is that Vedic calculation is available in toto, and it has been accorded some respect. Had Veda given figures in approximations, it would have lost all its legs to stand upon?

The Vedas themselves leave no room for any misunderstanding. They state that a saṁvastsara is *triṇābhi* (triple- centred, i.e., elliptical). It is not a single-centered common circle. And yet it may be taken as single-centred one for the purposes of calculations on the basis of mean radii. Mathematics supports Veda on this point. Modern science is fully aware of the fact that our earth does not move at a uniform velocity in its orbit around the sun. Its velocity is maximum when it is nearer to the sun. This is known as the linear velocity of the earth. But its angular velocity is always the same. Angularly it crosses one degree distance during each day-night unit (not the rotational day-night). Calculations on the basis of the elliptical centres may be tough. The mathematical basis of the angular completion of a circle at 360^0 is a far easier method. Thus Veda itself has stated the method of calculating *saṁvatsara*, solar, lunar years and the adjustment of their respective periods with the year. The very basis of calculation being 360^0 degrees makes it scientific in all certainty.

What happens to the time when *mahāpralaya*, great dissolution, occurs? Does the time also get dissolved like all forms of matter and anti-matter? The Sāṅkhya has not enumerated time in its list of 25 elements, as time makes no contribution towards creation. Time is not a creation of *Prakṛti*, the energy, either.

We quote below from a paper read by Dr. Mahāvīra of New Delhi at a seminar at Ajmer (India) held in October, 2000. It refers to the thesis of Dr. Stephen Hawking in his book the *Story of Creation: From Big Bang to Big Crunch* that time survives. Let us read Dr. Mahāvīra:

"The four *sūktas*, wise sayings, namely, R.V. 10-129/130/154 and 190 are famous under the name *Bhāvavṛittam* because in all of the four sūktas the state of creation of universe has been narrated, hence the *four* are called *sūktas on creation*. It is amazing that the name of Stephen Hawking's book on the subject of the science of creation which happens to be *Story of Creation* is a translation of *Bhāvavṛttam* itself. Be it as it is ! The first *Bhāvavṛttam sūkta* of Ṛgveda, i.e., 129th *sūkta* of the 10th *maṇḍala*, which is also known by the name *nāsadīya sūkta* begins with these words:-**nāsadasīnno sadāsīttadānīm.......** In this first quarter of the *mantra* the state of dissolution, which precedes creation, has been narrated in which it is stated that during the state of dissolution both *sat* and *asat* were not existing. It is noteworthy here that both the Vedic terms *sat* and *asat* have been used to denote highly scientific technical meaning to explain which there is no room here. What we want to specify here is that in that state of dissolution, which has been narrated as *sat* and *asat* being non-existent, the existence of one element has been clearly admitted, and that is time which has been specified by *tadānīm*. It means at that time when even *sat* and *asat* were not existing *kāla* (time) was very much in existence. The thesis of Stephen Hawking and Einstein regarding time have been knocked down by Veda with a very simple and straightforward word *tadānīm*. The Vedic scholars have so far not paid their attention to the scientific explanation of all of *Bhāvavṛttam sūktas* and that of this particular *mantra*. The existence of time during the state of dissolution and the calculation of time during that state, which is a serious scientific challenge, have been dealt with at length in Indian *darśana* books. During the state of dissolution, with the non-existence of sun, what should be the measure or basis of calculation is a major problem of science. But the ancient Indian *ṛṣis* have solved this riddle successfully in a scientific manner. The famous modern astrophysicist Karl Sagan has confirmed this fact by admitting the same in highly appreciative words in his interview which appeared in the issue of the *Hindustan Times* dated 26th January, 1996".

It may further be added that *Vaiśeṣika* treatise states, *kāla* (time) to be an eternal element. During the period of *pralaya* or dissolution it remains dormant and reappears in a state that is measurable during the period of creation for it provides the basis for measuring all sorts of activities let loose in the realm of relativity, and also serves as an instrument to human beings or any such form of highly intelligent species emerging therein to learn and verify the divine knowledge.

— 20 —

PRALAYA

Veda states *pralaya* (dissolution or recollapse) as an inevitable end of creation. *Pralaya* is inevitable in the same manner as creation is inevitable after *pralaya*. According to Veda that is the course of *Prakṛti*. Veda does not admit universe or creation to be infinite or eternal. It does not exist for ever. Once the process of creation gets initiated, its end is certain. Then again the end in itself is not final. It is also a phase which terminates with the emergence of creation again. Veda calls the period beginning with the commencement of creation, its growth and development and its final culmination a *day of Brahmā* or a *kalpa*, and the period during which the process of creation is not in existence, a *Brāhma night*, as well, is equal to it and also called a *kalpa*. The distinctive feature of *Brāhma day* is light while darkness is said to prevail during its night.

Every object or image in the universe gets created, and whatever is created comes to be erased. This rule is universal in application without any exception whatsoever. But the matter that goes into the composition of an object, does not get destroyed even after the destruction of the object itself. The matter gets transformed and continues to exist. In fact matter in itself is but condensed energy. The energy is neither born, nor destroyed. It is in eternal existence. As matter is a manifestation of energy, that is why its Vedic noun is *satya*. This aspect has already been discussed earlier. Merely the form or image in which matter manifests itself gets destroyed. It has also been pointed out earlier that the Sanskrit word *nāśa* means to become invisible or to disappear and not to get annihilated. In this respect Modern Physics states conservation of matter at least at atomic level, but Veda declares a complete dissolution back into energy during *Mahā-Pralaya*.

Veda even goes beyond modern science. According to Veda the existence of matter has a limit and when the same is reached, matter too, comes to an end. In other words, matter has its bounds. Matter has resulted from a transformation of energy. Its dissolution merges it back into its original state, i.e., energy. The state of matter is not altogether indestructible. It finds its end in regaining its origin through a cyclic transformation. Modern science accepts this position only partially. It agrees that the meeting of a matter and an anti-matter particle does bring about total annihilation of the two with a 100% transformation of their mass into energy. But, like Veda, it has not been able to take any such firm stand that the matter which arises from energy, finally merges back in the same. It admits the first half, but is quiet regarding the other part. At best it admits of a partial merger or dissolution to the original state.

Vedic wisdom states that universe is the result of the action born of the resolution of infinite conscious existence called Puruṣa that seeks expression in numerous forms through procreation. Such a resolution caused the emergence of rise in temperature which initiated motion in the vast ocean of energy at rest (*sattva*). The appearance of *rajas*, motion, was followed by *tamas* which wrought a part of *ṛta*, K.E., into matter through the process of condensation. *Ṛta* has a simple harmonic motion. Modern science endorses the final state narrated above. The motion continues in various frequencies. But Vedic wisdom reiterates that the creative frequency of *ṛta* culminates at the final leg of *mahāpralaya*.

The noun *pralaya* is derived from root *lī* to dissolve with the prefix pra and the suffix *a* added to it to form the said noun. Thus the *laya* part denotes dissolution in its original state of being. But prefix *pra* indicates that dissolution takes place as a part of an onward process. The noun *pralaya*, therefore, informs us that the way the creation of matter gets initiated and developed by the process of condensation of *ṛta*, similarly its final dissolution is also the part of the action of energy in motion. Creation comes to an end in dissolution. Prefix *pra* further indicates that onward motion continues resulting in re-emergence of creation. Energy that goes through creation and dissolution, neither increases nor decreases, it simply keeps on transforming itself while in motion. Veda names the said state of existence *avyaya* meaning a state of conservation, literally speaking non-spendable.

This position that *Prakṛti* as material cause, *upādāna-kāraṇa*, of universe, remains in eternal existence, it neither grows nor decays, and remains conserved as such, has been stated by Veda and Vedic school of thought ever since Veda appeared in the present form. Modern science in its quest has found such a position a fundamental truth of science. Matter is nothing but totally and truly energy condensed in form. Veda has emphatically stated this by naming it *satya*. This has been discussed at length earlier. Modern science had no such conception before. But the advent of Einstein established the fact that it is *satya* (real) in a manner that Modern Physics virtually occupied a Vedic pedestal non-chalantly.

The Vedic school has equally forcefully maintained that manifestation is a middle stage which is preceded and followed by unmanifested state of being. It has all along further maintained that manifestation or no manifestation, the conserved element is governed by an eternal motion. Add this up and the conclusion is inescapable that dissolution is but a completion of a round of the wheel of time to be followed by another round of creation and dissolution and so on eternally. And Veda has all along been stating science, but modern science has come to its threshold only in early 20th century of Christian Era.

We find that matter has further blossomed into organic matter resulting in the growth and development of a wide variety of organisms upon our planet earth, the *homo sapiens* being by far the most intelligent among them. Veda has classified a group of celestial bodies on the basis of there being available upon them certain conditions which render them habitable for life. Such a class of planets is called *vasu*, a term derived from the root *vas* to dwell with the suffix *u* added to it. It denotes any such planet in any galaxy which is suitable as a dwelling for living organism. Veda does not speak a word supporting exclusivity of earth as a life-bearing planet. On the contrary Veda maintains there being numerous planets in universe which are abodes of life. But modern science has only lately come to be kind to such a view, as a number of space radio signals have regularly been beeped, so far in vain, in the expectation of any response from any intelligent life in the universe. Even though no definite evidence has come to light, yet certain planets having conditions similar to our earth have been picked up in nearby galaxies and the spacecrafts have also sent in some supporting evidence from other bodies of our own solar system. It appears that science is going to accept Veda sooner than later.

It is very heartening to see modern science taking new and longer strides in its endeavour to reveal the mysteries of universe. We are here at the moment trying to present what the oldest book in the human library has been saying regarding the science of physics. The subject under focus is dissolution or end of the universe. But before we proceed further, let us glance through what modern science, equipped with the most powerful instruments in history and associated with a number of colossal minds, has to say regarding the future.

Modern scientists have, after a detailed study of various aspects, estimated the life of our sun to be between 8 to 9 billion years. The sun is no extraodinary star. It is rather a medium-sized one. The life of a star depends upon the period the hydrogen content in its body takes to burn itself out. According to scientific estimates about 90 per cent of our sun's life will be consumed in such a burning process and this sun would continue to scatter light and heat during the same. It is already past half of its age. For the purpose of this discussion we presume its remaining life to be around 4.5 billion years. (In the earlier chapter on kāla, time, our Vedic calendar related a *Brāhma* day-night period to be of 8.64 billion years duration.) Out of the 9 billion approximately 4.5 billion years have passed since preceding mid-night point. The part of day time remaining since dawn about 2 billion years ago is almost 2.35 billion years. If this is deducted from the life of sun, remaining 2.15 billion ycars would bc over before the present *Brahmā* reaches its coming midnight.

Let us take a step backwards. The sun has completed half of its life. The present day of *Brahmā* began some 2 billion years ago. So our sun was over 2 billion years old when it was dawn for *Brahmā*'. This places time of the birth of our sun about the midnight point of Brahmā's pre-dawn night. To cut a long story short, the life of our sun approximates *Brahmā's one day* and two half nights, or a day and a night period. In astronomical terms 8.64 billion years of *Brahmā* and 8.5 billion years, mean of 8 to 9 billion years of the life of our sun, are almost equal. Is it that Veda in its own inimitable way is giving us the life span of a medium-sized star like our sun?

This also reveals another side of dissolution. The life on our planet or anywhere else in this solar system would come to an end some time before the sun burns out its hydrogen fuel. That line appears to be drawn around the time *Brahmā* completes its day. The death of our sun will virtually be sunset for *Brahmā*. This means that it will be the end of biological life in this part of the universe and maybe, in some other planetary systems of our galaxy as well as in some other galaxies, which have a medium-sized star as their sun, and which also came to take birth in its own part of heavens close to the time our sun was being born. In this discussion *Brahmā* qualifies a periodicity.

According to some of the modern physicists the universe was born about 15-16 billion years ago from a big bang. Some others assess this period to be about 10-12 billion years. The period of a *Brāhma* day-night is 8.64 billion years. If a period of 2 billion years is added it comes to 10.64 or about 11 billion years which is the mean figure between the scientific assessment of 10-12 billion years. Even the assessment of 15-16 billion years falls short of a total of two *Brāhma* day+night periods, which add upto 17 billion plus years.

The fact is that modern science has been increasing its estimates regarding the period the universe was born based upon the distance it finds a distant ray of light to have travelled before reaching our earth. But lately it has been able to develop, more sensitive ears than its eyes peeping through the depths of vast expanses. A radio-telescope has enabled science to hear from considerably greater distances. Yet the scientists have so far not succeeded in locating the end of universe that they are searching. Recently they have been successful in establishing a telescope in the space around the earth beyond its atmosphere increasing their peeping capacity. Sending spacecrafts to considerable distances away from the earth is also a step in the same direction. That is why, probably, science has now changed its language and is speaking of the expanse of the known universe.

In ancient India Maharṣi Bharadvāja had completed a book named *Yantra Sarvasva* (Total Mechanics). It had 40 sections, one of which was related to aircraft. No copy of the said book is available but a part of the commentary written by Bodhānanda Yati on the said aircraft section could be traced by a Sanskrit scholar in pre-independence India. In it there is a hint of journeys from earth to mahaloka. This is Vedic noun for the inner galactic space in our galaxy, the Milky Way. It commences beyond the bounds of our solar system. Modern space-crafts are still contemplating to cross the said limit with manless space-voyagers. But even in such an exclusive commentary any reference to the end of universe was not suggested.

Out of whatever ascertainments have been presented by the researchers in the field of Modern Physics about the origin of universe, big bang theory has been credited most. According to the same, in early stages of creation an enormous primordial fireball exploded scattering its contents in all directions of space around. The force of explosion propelled the exploded radiation matter. The force of gravitation secured formation of billions and trillions of proto-stars in inter-space among its enormous chunks and hydrogen clouds. Gravity caused the body of proto-stars to shrink and the shrinkage caused rise in condensation as well as in its temperature. In tens of million years temperature fell to the state of commencement of a fusion process.

Our sun is a medium-sized star. Like our sun the burning hydrogen matter in its body keeps an ordinary star alive most of the period of its life. When its atomic fuel gets exhausted and burning begins to cease, the star heads towards its death. Modern science is conducting research study in the depths of heavens and has amassed a great amount of knowledge about the advanced age and final demise of stars. But it is still not certain, as Veda is, about great dissolution or any other state at the end.

In the life of galaxies and stars the basic factor is burning of atomic fuel. The fourth *mahābhūta Jala* influences the excitement generated by the third *mahābhūta Agni* to fuse excited matter into isotopes or atoms. This process continues as long as hydrogen lasts burning. Each stage of higher atomic formation requires specific temperature to enable

Jala to fuse the same into specific atomic moulds. The hydrogen fuel goes on increasing temperature under the force of gravity. But as and when hydrogen gets burnt up in toto rise in temperature gets arrested unless the helium matter which has been fused by burnt up hydrogen, starts to burn to keep temperature rising.

Fusion of hydrogen matter into helium starts when hydrogen burning reaches a temperature of 10 million K. Helium is fuel for the next stage which commences when rising temperature crosses 100 million K. effecting helium to burn causing fusion of carbon matter. This is the fuel for next stage. The carbon flash requires a temperature of 600 million K. which is reached after helium keeps burning for billions of years. Also at each stage burning material being far more condensed the bulk of star under gravitational pressure becomes comparatively less than the earlier one.

It is obvious that a star, that does not possess enough hydrogen matter to generate sufficient helium fuel to last till carbon flash point is reached will not be able to journey further. Our sun falls short in this category. Modern science says it will not be able to reach the carbon burning stage, its hydrogen being not sufficient to produce enough helium fuel to last that long. Its helium burning would transform it into a comparatively smaller-sized brightly shining white star called a white dwarf, which will lose its shine as it dies with the exhaustion of its burning material. As stated earlier, the life of our sun is estimated around 8-9 billion years by modern scientists.

But there are stars considerably bigger than our sun, having sufficient hydrogen fuel to burn and produce helium fuel sufficient enough to effect carbon flash which initiates carbon burning. This comes to an end at the fusion of iron matter. As the exoergic atoms end at iron, the atoms that succeed it on the periodic table are known as endoergic, i.e., needing more energy for fusion at the end than provided by the relevant reactants. Under such a state no amount of matter burning would emit any energy. With the absence of energy emission the gravitational pressure makes a short shrift of the bulk of star which suddenly gets squirmed and then like a spring released from pressure, in a tremendous explosion re-emerges in the space scattering its matter all around. Modern science calls such an explosive star a supernova, which becomes hundreds of times brighter than our sun.

It may be recalled that R.V. 1/163/9 describes the supernova event in its brief narration of the creation of universe by the galloping radiation in Chapter 8. In reactions generated in a supernova explosion all other atoms, that are available in nature, get created. These include all the numbers shown in the periodic table from iron onwards to No. 92. Scattered in space in the manner stated above its atoms getting mixed up with star dust or hydrogen clouds become parts of the new generation of stars and planets.

As already stated, only a few of the bigger stars having sufficient fuel stock within their respective bulk, succeed in reaching supernova stage. Others, so to say, fall on the wayside during this jounery in time. Those having insufficient hydrogen fuel, may not be able to reach even the helium flash. Many among those which pass this point, may succumb for want of more helium fuel at any point of time during the interregnum between helium flash and carbon flash. There may still be quite a few which fail to reach the iron manifestation point. Modern astronomers have categorised such failed stars or those that would fail and named them on the basis of some specific characteristics.

The supernova explosions occur in course of time. A few were observed during the past from our earth and the event was recorded. The last supernova explosion, which occurred within our own galaxy, the Milky Way, was seen in the year 1604 of Christian Era. They have been observed in other galaxies as well.

The journey does not end, according to modern science, with supernova explosion. But it takes a turn. The tremendous pressure of gravity, in the absence of the counterbalancing pressure generated by energy emission disappearing suddenly, compresses the core of star to the extent of transforming it into neutron matter. It means all the space within atoms having been expelled by fusion. Naturally the neutron matter is heavier than any stretch of imagination.

Even this neutron body continues to be further compressed under the force of gravity. Although the gravity is very strong, yet the rays of light still escape from such a body. But the increasing gravitational force at this point bends them considerably. Continuous compression keeps on shrinking the size of neutron star and mounting gravity succeeds in forcing horizontal rays back to body. This is the preface stage of a black hole, i.e., only the light rays that pass through an event cone, are able to escape, those touching even the perimeter of the cone, fall back to the body of the star. But contraction goes on and as the body surface gets squirmed to a critical radius from its centre, nothing is able to escape, a black hole comes into being, and everything gets trapped inside it. The surface radius at this point has been named event horizon.

Within a black hole the body of the star could continue to shrink further but event horizon does not change because gravity depends upon the distance from its centre which remains unchanged. In other words, the perimeter of a black hole is the critical distance from its centre at which the force of gravity does not let even rays of light to escape it. If the surface of star has further shrunk inside, it does not matter.

As no light reflects back from a black hole, modern science is still at a loss to verify the state of black holes scientifically in universe. Scientists are investigating the possibility through observation of binary star systems which disclose X-ray source. In the meantime black holes are a subject of lot of speculations among modern scientific world. Any matter that gets directed towards a black hole is lost forever. The capacity of a black hole to gobble up matter beats all speculations. The matter gone into a black hole simply turns into matter lost. It disappears from all accounts of matter maintained or drawn up by modern scientists. It ought to increase the bulk of matter within the black hole. There is no such evidence. The scientists are unanimous that consumption of matter by a black hole does not conform to the pattern of consumption laid down by science. That is why some speculate black holes to be some sort of cleaners. They may be working as gravity cleaners of universe. Some even presumed that matter gobbled up gets flushed out in some other universe for they fail to accept black hole as the ultimate end. Modern science stands marking time at this juncture till additional evidence comes along.

It is clear from this study of the life of stellar world which we have briefly narrated above, that modern science has still not been able to clearly reveal either the beginning or the end of the universe. It enumerates a lot of objects it found in nautre, but it is not clear as to how elementary bodies came to be formed. So stating the properties of such elementary particles, etc., it starts explaining how Creation started with a primordial *bang* initiating the process of fusion of hydrogen matter into the large family of atoms. Similarly it describes stellar journey through space and time, but nearing the end it lacks clarity once again. While descriling Black Hole phenomena it forthrightly admits that its rules regarding consumption of matter do not set well with its study of Black Holes. Modern science also fails to show as to how inertial energy was able to create such a superbly organised system as we find our universe to be.

In mentioning all this, we have no intention of attempting to present Modern Physics/Astrophysics and the most laudatory achievements given to us by modern science which have revolutionised our life itself in any less praiseworthy light. Our purpose is to emphasize that the journey of science is always and without exception a journey in search of truth and modern scientists have carried the banner of truth without a doubt even at the cost of being persecuted rather than rewarded. And the finale of any journey in search of truth must be the realisation of the final truth. The ancient Indian scholars, known as *ṛṣis* have handed down Veda as knowledge revealed. Throughout their emphasis is on the realisation of truth and nothing else but the truth.

Modern science is in search of truth and Vedas are given the credit of stating truth. Hence the two lines must converge and must try to convey the same meaning, maybe in non-identical terms and language. We have nothing new to add to Veda nor we can change a single consonant of the Vedic text. The best we can do is to try to interpret it in the manner prescribed by scholars of highest order. This narration related to modern science is to locate the direction and the points of such convergence. It will be a test of the theory of Vedic revelation as well, for there is no other supporting evidence by way of observation and experimentation. If the convergence is sufficient, and whatever else is being stated by Veda regarding areas where modern science is still groping, appears to be plauṣible, we are sure that modern scientists would be the first to take a closer look at Vedic knowledge.

Let us now ponder over what Vedic wisdom has to say. Earlier in this chapter *jīvapralaya*, bio-annihilation, yet to occur on this earth of ours, has been explained in terms of a *Brāhma* day and the life of our sun. It has been pointed out there that the period of a day-night unit of *Brahmā* as stated by Veda and the life of our sun as assessed by modern-science approximate each other and our sun being a medium-sized star, that may be a Vedic way of mentioning the approximate life span of such a star, which has been used as a standard of measure for time-distance upon this earth. As the downing of our sun meant disappearance of the source of light in this part of heavens resulting in deepening of darkness here, and maybe elsewhere as well in the universe, with the lives of a few other contemporary medium-sized sun-like stars also coming to be extinguished around that time or in quick succession within the span of 4.32 billion years, the period of so-called *Brāhma* night becomes a synonym of *pralaya*. Veda being revelation of knowledge upon this earth for the benefit of earthlings the adaptation of life and events here make more sense. This day and night cycle of '*Bramhā*' keeps on rolling and stellar deaths, as and when occurring in time as per the studies conducted by modern science as briefly narrated above, the *khaṇḍa pralaya*, partial annihilation, goes on happening in the galactic expanses at various points of time at regular intervals.

A few words about universal life span may not sound out of place here even though it has earlier been mentioned in the chapter on *kāla* (time). According to Vedic seers the age of the present universe is about one third of a quadrillion human years, 50% of which has passed. The state of *mahāpralaya*, great dissolution, will come after the other half is completed. The said chapter on *kāla* ends with a quote from Dr. Mahāvīra of New Delhi who has pointed out that the title of the book *Story of Creation* by Dr. Shephen Hawking happens to be a virtual translation of the *devatā*, i.e., subject matter, of four Vedic *sūktas* from the 10th Maṇḍala of R.V., namely, *bhāvavṛttam*. We agree with Dr. Mahāvīra, but would like to say that the Vedic term *bhāvavṛttam* could as well be translated as a *A Biography of Brahmā, the Atom*, for the story of the universe or creation is a narration of such a biography.

Modern science believes to have located a new standard to measure astronomical distances and time in the velocity of light. On that basis it has calculated the distance light travels in one human year, and standardized the same. It has also been subdivided into smaller durations. A.V. 19/53/1 with its *devatā* as *kāla*, time, specifically uses the term *saptaraśmi* describing the ray of light as a seven colour braided measuring cord of time. The term *raśmi* has a dual meaning, a ray of light as well as a measuring cord. But the exactitude in determinations achieved through *saṁvatsara* system appears the reason for Vedic preference for the latter. Veda has standardised the duration of its day and night as a standard unit of measure for establishing dimensions of universe and time both. It has all of its calculations worked out on the basis of *saṁvatsara* which is fundamental to all such determinations. A day/night period of *Brahmā* is subdivided into 1000 units and each unit further into 10 sub-divisions which conform with the *saṁvatsara* of the planetary mass of our sun completing one orbit around its procreator. This has earlier been named *kaliyuga* meaning No.1 unit. It has further been sub-divided in relation to the *saṁvatsara* period of the earth around the sun and so on. The details have already been given earlier. Here we are discussing *mahāpralaya* which brings the entire sweep of creation in time under consideration. For this *Brahmā*'s day and/or night unit is sufficient.

According to Vedic seers the life of our *Brahmā*, i.e., the atom of our present universe is to run 36000 day-night units or 100 years determined on the basis of 360 per count. This when calculated comes to about one third of a quadrillion years in human terms. In other words, the matter in atomic formation is to survive that long in present Creation. Under such a dispensation of the creative process how can atomic form be annihilated earlier save two exceptions, a 100% annihilation when matter particle meets its anti-matter particle, and a partial loss when matter particles make readjustments at every fusion. The two exceptions are extremely limited in nature and provide no clue to ultimate total annihilation or dissolution of matter call whatever you like. But the two exceptions establish the fact without a shadow of doubt that matter, given the right conditions, is capable of changing itself back to energy.

Veda asserts that the process of creation finally ends up in dissolution of all matter. Veda further asserts that each creation is followed by its dissolution and each dissolution finds its end in a new creation. They are following each other in an infinite cycle. It may also be deduced from whatever little has been left available in ancient literature that the duration of the period may vary; it may not be uniform, but the present creation is to last the period already mentioned. It has been stated that in some of the creations prior to the present one, *Brahmā* possessed five faces rather than four as is the case of present one. *Brahmā* has been analysed and interpreted in detail earlier to establish that it is a Vedic synonym of atom, and its four faces are four sub-orbitals. What has been referred to above regarding the existence of a five-faced *Brahmā* in one or more earlier creations simply means the birth of atoms having five sub-orbitals in that universe. Of course, such inferences, or even conjectures, result from nothing but scientific thinking.

The description as recorded says that one Maharṣi Mārkaṇḍeya saw earlier manifestations of *Brahmā* wherein he saw a few five-faced ones. In Sanskrit language seeing also means knowledge, realisation. Maharṣi Mārkaṇḍeya belongs to Vedic age in India. He might be observing and studying the orbital motion of an electron in the atomic composition. He was a scientist of his age. Obviously he knew that the velocity of radiation energy generated by *Hiraṇyagarbha* explosion is the cause behind propulsion of electron which determines its sub-orbital position. He might have calculated that present energy of universe just exceeds four sub-orbitals, but falls short of making a fifth one. The record says that generation of a fifth sub-orbital of our *Brahmā*, the present atom, failed to reach such a state and the process got cut short to four sub-orbitals only. He might have realised that excess explosive release of energy was responsible for the number of sub-orbitals and such a phenomenon of excess energy release occurring from *Hiraṇyagarbha*, primordial explosion, could result in manifestation of the composition of *Brahmā* having increased number of its faces, sub-orbitals.

We may also recollect correct choice of words and expressions in Veda denoting scientifically correct positions or meanings which stand thoroughly verified by Modern Physics, more so as the latter had no regard of any kind whatsoever for Veda as an authentic source of scientific knowledge. The all-encompassing Darwinian influence was extremely ill at ease even to listen to, not to say of accepting, any suggestion that an antique book like Veda could be an expression of infallible knowledge. Even then, experimentation apart, we, in all humility, make ourselves bold enough to write that Modern Physics, and even Astrophysics, has not so far discovered or established anything worthwhile new that has already not been stated by Veda. It is a different matter that scientific community was by itself not in a position to understand and interpret Veda on their own, and certain Western scholars, who in fact put in superhuman endeavour to know Veda, were most unfortunately overwhelmed by motives other than any search for truth. Veda does not contain anything for India in particular, or for that purpose for any race or country, save the human race and the mother earth. In spite of all this Vedas do not say that you accept them as a matter of blind faith. Their entire emphasis is upon knowing the truth and nothing but the truth. Gold never shies at its veracity being tested a thousand times.

While discussing the state of universe we would like to refute a grossly misunderstood position related to Veda about the duration of its existence. It is utterly wrong to say that Veda put this universe to be about 2 billion years old. We have already explained this aspect that this approximation is about the present day of *Brahmā* and certainly not regarding this cosmos. Of course a day indicates activity of creation and a night rest or *pralaya*. With regard to creation culminating into biological development upon earth, it did get initiated with *Brahmā*'s present day, and would end at its sunset so far as our earth is concerned within the specified duration and to that extent it is the beginning of creation for earthlings. But earlier it has been reiterated for the sake of clarity and correct understanding that the creation of the present universe began about one sixth of a quadrillion human years ago, that modern science still falls far too short of it, and that universe is yet to run almost the same length of time. The latter position accommodates even the end of such stars which are estimated to last trillions of years by present scientific community. Veda is yet to be proved wrong.

Again the great dissolution stated by Veda is to take place only at the fag end. Before that matter survives in its atomic form all sorts of destruction, barring the two exceptions mentioned earlier. It is not a very old issue when modern science, having discovered anti-matter and its rather small presence in this matter world, thought of the existence of a separate anti-matter universe somewhere in a distant part of heavens, and that it imagined the state of final annihilation by the two matter and anti-matter universes coming together at some point of time in future. But subsequently emphasis on such a possibility appears to have lost force. The present state of thinking is due to indecision for want of any further knowledge or clue. Modern science appears to have preferred to state to the extent it knows and to keep mum regarding the rest.

But Veda has no such hesitation. That is why it has clearly and unambiguously stated that the cycle of creation gets completed with return to energy by way of final dissolution of all matter and very significantly, anti-matter as well. We have not been able to find any details regarding such a change, except that the same has been stated in no uncretain terms. That is why there is no opportunity to explain the same by way of expounding the derivation of terms used. The nouns *pralaya* or *mahāpralaya* have already been explained, but they simply communicate the state of change from matter to energy, not how. Whatever little indication has caught our attention will be elaborated below. But we feel that Veda has at least a fairly convincing *prima facie* case in this regard and we would like to enumerate the grounds:

1. We find that any consistent motion in material universe has the tendency to move in a spiraling pattern in elliptical, parabolical or hyperbolical fashion under the influence of force. Anybody under motion, after completing its round, returns almost to the position it commenced its journey from, not quite, but somewhat way ahead.

Veda has named such a round a *saṁvatsara*. The journey of creative activity began with energy at rest becoming kinetic. All development is the effect of five *mahābhūtas* which by themselves disappear in the state of *mahāpralaya*. Hence the run by energy, which went kinetic, ought to return at the conclusion of final *saṁvatsara* to a similar position, of course, way beyond the point in time it commenced its journey from (like a wave).

2. The very nature of all creative processes in universe is averse to leaving any waste material. There is no waste in universal process. The narration of the journey of stars in time as visualised by modern science and summed up briefly above, fails to take us beyond an enormous graveyard of numerous hulks of stars and black holes. Such wasted stars are an antitheisis to no waste efficiency. Dissolution is the most efficient disposal of waste in tune with the law of conservation as well as readiness for a *de novo* creation.
3. Nobody knows what happens to matter gone into a black hole. First, for some even the existence of a black hole is not a certainty, but quite a bit of circumstantial evidence impinges upon the minds of modern scientists to think about such a possibility. The existence of a neutron star after a supernova explosion is verifiable. After further gravitational compression the state of star at which light can escape through event cone is verifiable, because such a reflecting back light is the direct source of gaining knowledge about the star nearing a black hole state. It is only when the body gets compressed to the state called event horizon that light ceases to escape its gravitational grip and the non-availability of any knowledge further through escaping light forces the scientists to presume that a black hole has come into being. Hence nobody knows what happens there thence onwards because the squirming star body may shrink further comparatively shorter to the event horizon. All that could have been made out is that matter being gobbled up in a black hole is lost; no more traceable. But any such matter being genuine in energy must remain conserved, if not as matter than as energy.
4. What is the zenith of the force of gravity? It appears that gravity

reaches its zenith or near about within a black hole. At neutron matter stage no inner space remains within an atom which may yield to gravity, yet neutron matter gets compressed which is clear from the black hole story. It gets compressed beyond event horizon. What happens when there is no more room to yield and the pressure goes on building in inverse square proportion?

5. By the time black hole stage is reached at least three of the five *mahābhūtas*, namely, the fifth one *Pṛthivī*, the fourth one *Jala* and the third one *Agni*, themselves disappear. The fifth one works the formation of molecules and atomic compounds. With the helium flash followed by the carbon flash the performance of *Pṛthivī* appears to become redundant. The fourth *Jala* and the third *Agni* are in harness till the supernova explosion takes place. The remaining neutron matter body does not leave any room for gravity to cause generation of friction resulting in temperature rise on account of heat created. Hence *Agni* disappears alongwith *Jala*. Only a neutron star body remains at the mercy of gravitational force. This continues upto the black hole stage. The *Jala* and *Agni* having bidden their adieu after supernova, only hurdle to be jumped before touching the tape of *mahāpralaya* so to say, is the second *mahābhūta Vāyu* as gravity, for the first *mahābhūta Ākāśa* is merely a field having no material properties save light. It ceases to exist with the cessation of light and motion.

6. We may as well consider the slight indication given by Veda. It is contained in the opening words of the first *mantra* of R.V. 10/129. The whole *mantra* with two more would be presented later on in this chapter as they describe the state of *mahāpralaya*. Here our purpose is to point out only the indication contained therein as mentioned above. The opening first quarter says: **nāsadāsīnnosadāsīttadānīm**. It means at that time (of *mahāpralaya*) there was neither matter nor anti-matter. This part of the *mantra* is also included in the quote from Dr. Mahāvīra's paper given at the end of the last chapter on time. But there the emphasis is on time factor. The mention here is to draw attention to disappearance of both matter and anti-matter in *mahāpralaya*.

Modern Physics endorses the fact that matter and anti-matter both annihilate each other which result in a 100% transformation of their mass into energy. In other words, the two dissolve totally in energy.

7. The availability of approximately 10% anti-matter in our 90% matter universe may cause to dissolve 10% matter. Still 80% remains which will end up either in hulks of the dead stellar bodies or the black holes. The latter may ultimately be able to gobble up all the hulks strewn in *ākāśa*, space, thereby collecting the entire waste in those gravity cleaners. That is why our question regarding the effect or the interaction of gravity at its zenith, a phenomenon which may be called *parama gurutva*, absolute gravity, in Sanskrit. We do not know. The Veda does not appear to say more. It is now upto Modern Physics to find an answer. The question may be worded differently to make it more direct. Does the state of supreme gravity ruling within a black hole foment suicidal reaction in matter within its realm? Of course, there remains black matter to be disposed of. The answer if and when it comes in the affirmative may virtually convert Modern Physics to be a copy of Vedic Physics. Otherwise the search for truth would continue.

8. It has earlier been pointed out that an electron orbiting around a proton is the smallest *saṁvatsara.* On the other extreme matter propelled by primordial *bang* finally completes the biggest *saṁvatsara* of creation when it finally converges and seeks transformation back to energy at the threshold of the *mahāpralaya.* The final dissolution leaves no scope for *Vāyu*, force, and the energy coming to rest makes *ākāśa* to disappear.

9. And last but certainly not the least is *Puruṣa* factor enumerated last by Sāṅkhya. *Puruṣa* as discussed in an earlier chapter is the efficient cause of universe as *Prakṛti*, energy, is the material cause. It has been pointed out that inert energy of modern science is in no way by itself capable of making a single move. A factor other than the same is incumbent to make it kinetic. Similarly the inert energy is as well utterly incapable to cause the creation of a superb orderly system like our universe. It requires the existence of a supremely

intelligent consciousness for providing proper direction. Such is the Vedic *Puruṣa* which pervades and over-reaches as well as over-covers the inert Prakṛti. It is present everywhere. If it can, so to say, switch on the creative process; it can equally efficiently switch it off.

The use of expressions like *switch on* and *switch off* above must not be mistaken to mean a beginning or an end of motion or activity in respective contexts. They simply communicate two location points of two ends of a phase in eternal cyclic motion. A synonym of *Puruṣa* or *Saccidānanda* given by Veda is *Paramātmā* which denotes an Absolute Being or existence in eternal motion. Like the existence of temperature under absolute zero, a dynamic state exists eternally beyond the level of perception, which is caused to rise to perceptiable intensity levels at the end of the periodicity of *mahā-pralaya*, great dissolution, whereby another universe starts getting manifested anew initiating successive round in time.

It would be of interest to end this chapter with a glimpse of the picture worded by Veda of the state of *mahāpralaya* occurring before the initiation of the creation of our universe. It is R.V. 10/129. The entire *sūkta* consists of seven *mantras*. We are just quoting the first three with their explanations below:

nāsadāsīnnosadāsīttadānīṁ nāsīdrajo no vyomāparo yat
kimāvarīvaḥ kuhakasya śarmannambhaḥ
kimāsīdgahanaṁ gabhīram
(R.V. 10/129/1)

(tadānīṁ nāsadāsīt) At that time of *mahāpralaya*, i.e., the great dissolution, there was no anti-matter **(no sadāsīt)** nor there was any matter. **(nāsīdrajaḥ)** There was no *rajas* that is motion. **(no vyomāparo yet)** there was no *ākāśa* or space either. **(kimāvarīvaḥ kuhakasya śarmannambhaḥ kimāsīt gahanaṁ gabhīram)** What to say of a cover, nowhere was even any shelter, no liquids, no solids; it was all inscrutable.

By denying the presence of matter as well as anti-matter in any form and also denying any kind of motion or space, liquid or solid,

Veda says it was an inscrutable state of existence. The *mantra* without naming, is emphasising the existence of *Prakṛti* (energy) at rest in juxtaposition with *Puruṣa*.

na mṛtyurāsīdamṛtaṁ na tarhi na rātryā ahna āsītpraketaḥ
ānīdavātam svadhayā tadekaṁ tasmāddhānyanna paraḥ kiñcanāsa.
(R.V. 10/129/2)

(tarhi na mṛtyu āsīt na amṛtam) At that time there was neither death nor any life. **(na rātryā ahnaḥ praketaḥ āsīt)** There was no way of knowing day or night. **(tat ekam)** That solitary one **(svadhayā avātam ānīt)** with the support of its own being was present alive without any air. **(tasmāt paraḥ)** Beyond that One **(anyat kiñcan na āsa)** there was none else whatsoever.

The second *mantra* after refuting the presence of a few more phenomena positively affirms the live existence of the solitary Being which existed on its own without needing breath. The mention of live existence denotes solitary *Puruṣa*, the inert *Prakṛti* having no consciousness.

tama āsīttamasā gūḍhamagre apraketaṁ salilaṁ sarvamā idam
tucchyenābhvapihitaṁ yadāsīttapasastanmahinā jāyataikam
(R.V. 10/129/3)

(agre tamasā gūḍham tamaḥ āsīt) Prior to creation, *Prakṛti* was engulfed in dense darkness, **(idaṁ sarvam)** all of this universe **(apraketam)** was nowhere to be known, i.e., was unrecognizable, because it was **(salilam)** dissolved in its state of causation, i.e., *Prakṛti* **(yat apihitam āsīt)** which was covered **(tucchyena ābhu)** by a zero complete in all respects and devoid of any result-oriented attribute, eternally unchangeable Being. **(tat ekam)** That solitary one, i.e., *Prakṛti* **(tapasaḥ mahinā ajāyata)** developed for creation with the warmth of the glory of that Being.

The Being stated by an adjective like *tucchya* meaning *śūnya*, zero (A.K. 3/1/56) is a clear pointer in the *mantra* to what Sāṅkhya has enumerated as *Puruṣa*. That is why such an adjective, which gives the sense of a zero indicating completeness, is a synonym to the alternate name stated by Sāṅkhya as *Puruṣa*, the wholesome.

It is highly interesting to point out that Dr. Hawking, among the modern physicists, has already come out in favour of a zero state. This text bears his quotation elsewhere.

— 21 —

VEDIC MODEL

From the time it was believed that our earth was the centre of the universe and was located on the back of a tortoise, modern science of physics during the last three centuries has become far better acquainted with the universe and the position of this planet of ours in the same. The source of the emergence of this universe has also been understood as well as the process of development of stellar and planetary worlds. Modern Physics has unveiled a lot of secrets regarding the space and the cosmos within it. But modern astronomers and physicists are still probing and have not so far succeeded in finally determining the initial steps related to the manifestation of this greatest organisation called universe.

As more and more knowledge regarding matter and universe is being revealed and significant aspects of its character are yielding before scientific endeavour scientists are busy trying to correlate them in some unified model that may explain satisfactorily and possibly comprehensively the mathematical equation which is working and holding the universe the way it has materialised and is ticking. The major issues are still how this universe has come to originate and how it would come to an end, or else?

The first important model stated the beginning with a *big bang* and its end in a final recollapse or a big crunch. Alternately a steady state model was also floated but it did not respond duly to the evidence revealed by researches. The initial *big bang* theory stays in a modified form with a singularity replacing the earlier concept. The motion in universe is explained by way of two parallel lines. If the two finally converge, universe would come to an end. On the contrary, if they diverge, universe would continue *ad infinitum*. But the state of a singularity reached in a black hole added weight to a beginning from a singularity.

Yet scientists are busy seeking to resolve the theory of relativity and the quantum mechanics. It is being pointed out that during initial as well as late stages of the universe the effect of gravitation would be too strong to ignore quantum effects. Hence the state of the universe at a beginning or an end, if it does have one, continues to remain under a sign of interrogation. And if at all it does, what would be its nature of occurrence like?

Prof. Stephen Hawking in his book, *A Brief History of Time*, has examined various models propagated by worthy scientists to explain the beginning of universe with a bang or otherwise, but they all, even though appearing to meet the evidence presently available, fail to explain some vital aspects observed. Let us hear him as he concludes his analysis in his above-named book:

> "In the classical theory of gravity, which is based on real space-time, there are only two possible ways the universe can behave: either it has existed for an infinite time, or else it has a beginning at a singularity at some finite time in the past. In the quantum theory of gravity, on the other hand, a third possibility arises. Because one is using Euclidean space-time, in which the time direction is on the same footing as direction in space, it is possible for space-time to be finite in extent and yet to have no singularities that formed a boundary or edge. Space-time would be like the surface of the earth, only with two more dimensions. The surface of the earth is finite in extent but it does'nt have a boundary or edge : if you sail off into the sunset, you don't fall off the edge or run into a singularity........
>
> "If Euclidean space-time stretches back to infinite imaginary time, or else starts at a singularity in imaginary time, we have the same problem as in the classical theory of specifying the initial state of the universe: God may know how the universe began, but we cannot give any particular reason for it began one way rather than another.

> "On the other hand, the quantum theory of gravity has opened up a new possibility, in which there would be no need to specify the behaviour at the boundary. There would be no singularities at which the laws of science broke down and no edge of space-time at which one would have to appeal to God or some new law to set the boundary conditions for space-time. One could say: The boundary condition of the universe is that it has no boundary. The universe would be completely self-contained and not affected by anything outside itself. It would neither be created nor destroyed. It would just BE."

Prof. Hawking further emphasises that this idea that time and space should be finite without boundary is just a proposal it cannot be deduced from some other principle. He says that like any other scientific theory, its real test would be whether it makes predictions that agree with observations. It is highly significant to observe that Prof. Hawking by advancing such a proposal is veering round to what Veda says as to how the universe came into being and what its future is.

This brings us to the title of this chapter. But before we take it up, it may better be remembered that according to Veda *Puruṣa* is a scientific reality which performs only and only on scientific basis. It may also be realized, as has been tried in this presentation of Vedic Physics that It (*Puruṣa*) is the real source of direction to prompt idle energy to transform into motion or action, to emerge into a supreme orderly system wherein even each particle, not to say of bigger bodies, inherently behaves in a manner contributing to support the cosmic pattern. Unfortunately the so called wisest species on this earth, the human race, are the greatest culprits in abusing their existence in learning the ways of science to defeat the mandate of creation laid down by that Extraordinary Being who is making *Prakṛti*, energy to modern science, dance in the most spectacular and yet queerest but fascinating style. According to Veda the system of universe is a fully self-contained whole. It has everything that it needs, including all the prosperity and riches that may be desired, nay, even imagined. These are simply to be earned out of the system. Instead, we expect God to grant them only to praying individuals in token of such prayers. It has shown the way; it continues to enlighten us regarding the method. Build a collective, homogeneous, organised system to derive what you want out of the system that has built itself including yourself. Try to learn its methods, i.e., the science, and employ them for the collective benefit. This concept rules all the prayers stated in Vedic texts.

Let us return to the subject of the Vedic model which has been specifically stated in Atharva Veda at 10/8/29. The same has also been stated in the *Īśāvāśya Upaniṣad* in following words:

pūrṇamadaḥ pūrṇamidaṁ pūrṇāt pūrṇamudacyate
pūrṇasya pūrṇamādāya pūrṇamevāvaśiṣyate

The above quoted *Upaniṣad mantra* is very famous and is widely known to the students of Sanskrit language the world over, but its scientific import has not been realised. Let us consider what the *mantra* is communicating. First a literal translation:

(pūrṇamadaḥ) That is a self-contained whole; **(pūrṇamidam)** this is a self-contained whole; **(pūrṇāt)** from a self-contained whole, **(pūrṇam udacyate)** emerges a self-contained whole. **(pūrṇamādāya)** Having taken a self-contained whole **(pūrṇasya)** from a self-contained whole, **(pūrṇameva)** only the self-contained whole **(avaśiṣyate)** remains in balance.

The scholars interpret the sense of the *mantra* given above as a statement of a mathematical axiom. Agreeing with such a view it is humbly submitted that it is the model of creation that is being presented in the so called mathematical form. The key is the word *pūrṇa* which deserves some elaboration.

Pūrṇa: Literally means a self-contained complete whole. The first *pūrṇa* preceding an *adaḥ* denotes the self-contained complete infinite whole of existence. Existence is the first characteristic of *Saccidānanda/Puruṣa*. It is also an attribute of *Prakṛti* as its nature is conservation. The *Puruṣa* enjoys infinite existence as has been explained in the chapter on the same as well as in that of the supreme Master of Sciences. It is all wisdom and intelligence as well as consciousness and dynamism. It has no form but is an all-pervading infinite conscious existence. In a union with *Prakṛti* it turns the latter also into an infinity in existence. Such an infinity of the union of Puruṣa and *Prakṛti* is the self-contained complete whole as communicated by the first *pūrṇa* in the *mantra* under consideration.

The second *Pūrṇa* followed by *idam* denotes the cosmic system called universe which manifests in space-time relativity through the transformation of *Prakṛti*, energy, in motion. The transformation is initiated and sustained by the impetus generated by the dynamic contact of *Puruṣa* in union which also provides the necessary wisdom and scientific knowledge as well as efficiency so well reflected in the orderly emergence of this system of universe. This universe as a self-contained complete whole is the second *Pūrṇa* emerging from the first one as stated in the *mantra*. The universe as a self-contained complete whole represents a complete system emerging from within the infinity of *Puruṣa-Prakṛti* union, the primary *Pūrṇa* in existence. The other enjoys its own existence within that infinity of existence, but in the form of a self-contained complete material system based on the transformation of *Prakṛiti* in motion, so says the *mantra*.

Let us try to visualise this in mathematical terms, which would be as follows:

Efficient Cause *Puruṣa* (existence +)	=	∞	
Material Cause *Prakṛti* (energy)	=	10	
Union of *Prakṛti* and *Puruṣa*	=	10^{∞}	$= \infty$
Emergence of Universe	=	$10 / \infty$	$= 0$

The number 10 symbolises initial conservation. The series of digits from 1 to 9 emerges from 0 and remains conserved in the same at the completion of each run. The digital run also denotes a periodicity of occurrences. In fact number 10 is once zero, 20 is twice zero and so on. The conserved existence remains a zero neutralised from both negative and positive sides. But in manifestation as numbers it keeps on conserving itself at each run. Such a run would continue infinitely maintaining creation and dissolution. But the run would commence only when the dynamism of *Puruṣa* overpowers the inertia of *Prakṛti.* The process of creation can be worked out by a division: $10 \div \infty = 0$, that is, this conservation would occur after each run of manifestation when divided by existence infinity and each run would in itself remain a self-contained complete whole in a state conserved throughout.

This mathematical analysis also neutralises the observation made by Prof. Hawking as quoted above that if the boundary condition of the universe would be completely self-contained and not affected by anything outside itself, it would neither be created nor destroyed. It would just BE. On the contrary, the above analysis affirms the Vedic position that each manifestation or creation would keep on emerging as a self-contained completely whole system within the infinity of dynamic existence and submerging within the same in an infinite process of occurrences. Veda has been declaring it from the very first day as an infinite cycle of creation and dissolution. Being and disappearing would be an infinite process in infinity of time.

Let us also review the propagation by the Atharva Veda. It says:

pūrṇāt pūrṇamudacati pūrṇam pūrṇena sicyate
(A.V. 10/8/29)

It means: **(pūrṇamudacati)** a self-contained complete whole emerges **(pūrṇāt)** from a self-contained complete whole and **(pūrṇam)** the self-contained complete whole **(pūrṇena sicyate)** is sprinkled by the self-contained complete whole.

The Vedic statement is more abbreviated and still more significant. Its first half part quoted above states the same contents that have been communicated by the earlier *mantra* from *īśāvāsya upanishad* already discussed above. But the later half is no less significant. The material whole in the form of universe is being sprinkled by the cosmic rays coming from the self-contained complete whole. What a poetic way of describing the cosmic system! A.V. further signifies that the secondary self-contained whole, i.e., the universe is sustained by the primary self-contained infinite whole throughout by way of cosmic rays.

It may be recalled from the earlier chapter on *Puruṣa* that the first *mantra* from the *Puruṣa sūkta*.(R.V. 10.90) in its fourth and the last leg states:

attyatiṣṭhaddaśāṅgulam

In the part of the *mantra* quoted above, the existence of *Puruṣa* has been described by Veda as surpassing the limits of the ten causatives, i.e., five *tanmātras* and five *mahābhūtas*. The *tanmātras* represent what is known as the quantum mechanics and the *mahābhūtas* identify themselves with general relativity of Modern Physics. The state where there are space-time and quantum boundaries is overstood by the infinity of *Puruṣa*, so says each one of the four Vedas. What Prof. Hawking, after discussing various models is suggesting as a proposal, has been declared by all the Vedas from day one, specifically narrating it to be a state of existence beyond the quantum or relativity boundaries and yet occupied by *Puruṣa*.

Maharṣi Dayānanda Sarasvatī, the greatest Vedic scholar and social reformer of the modern age, has specifically so explained the above quoted part of the said *mantra* in his famous Introduction to the Commentary on Ṛgveda etc., written in the year 1876 of the Common Era. It runs as follows:

> **pañca sthūlabhūtāni pañca sūkṣmāṇi caitadubhayaṁ militvā daśāvayavākhyaṁ sakalaṁ jagadasti...... etasmāt bahirapi vyāptaḥ sannavasthitaḥ.**

This says: **(pañca)** five **(sthūlabhūtāni)** *mahābhūtas*, namely *ākāśa,* vāyu, agni, jala and prithivī covering the entire field of relativity **(ca)** and **(pañca)** five **(sūkṣmāṇi)** tanmātras covering the quantum state **(etadubhayam militvā)** adding these both **(daśāvayavākhyam)** said to be based upon ten organs **(sakalaṁ jagadasti)** is this entire universe .... **(etasmāt bahirapi vyāptaḥ)** pervading even beyond these **(sannavasthitaḥ)** and well established.

The above quotation from the said book of Maharṣi Dayānanda written in 1876, when the theory of relativity and the quantum physics were not known to Modern Science, shows that a Vedic scholar like Dayānanda was stating and explaining the matters connected with the birth of this universe in terms reached by modern science much later, and on the basis of Veda, commenting or elaborating aspects which still remain unclear as they are still being groped.

It may also be mentioned that the specific state which set the ball of creation rolling or flying with a bang has been called *Hiraṇyagarbha* by Veda. The *Hiraṇyagrabha* may be considered the singularity as the same, according to Veda, contained whatever was subsequently to take the shape of the universe we see. But for the singularity of *Hiraṇyagrabha* there was no second being. Consequently there was no sense of duality and perceptions that occur on account of there being more than one manifestation. There were *Ākāśa* the field, *Vāyu* the force, *Kāla* the time, *Dik* the direction, within that singularity in sort of an embryonic state and *Prakṛti* and *Puruṣa* both in a state of formless infinite existence. *Agni* was there likewise with temperature rising from under absolute zero to higher degrees building up to the point of big bang burst.

The *Puruṣa*, as per *Taittirīyopaniṣad*, desired to manifest as many by procreation. Its will caused *Agni* to rise from under absolute zero state. *Prakṛti* started pulsating which in turn resulted in stirrings. All such developments were confined within the singularity of *Hiraṇyagarbha*, which caused a highly condensed and concentrated manifestation of radiation and *Prajāpatis*, i.e., particles, both matter and anti-matter.

The part *mantra* **yathāpūrvamakalpayat** (R.V. 10/190/3) very clearly states that the present universe was created in the manner similar to the earlier ones, thereby affirming succession of universal creations. This further implies such succeeding creations followed by great dissolutions as a non-stop continuous process being caused by *Puruṣa* in union with *Prakṛti*. The mathematical presentation mentioned above corroborates such a conclusion that the creative process will continue ending in a naught, dissolution, each time in the infinite union of existence and conservation, i.e., *Puruṣa* and *Prakṛti*.

— 22 —

VEDIC CALENDAR

The entire accounting of different periodicities beyond eons occurring in time is based upon astronomical happenings with our earth at the bottom as its base. For earthlings that is the most effective, convenient and accurate method prescribed by Veda. No other calendar matches it. It employs astronomical method right upto the universal level and beyond to eternity. Our sun and earth together form a mighty clock, the sun, its dial and the earth, its pendulum, the major difference being in the swing. Unlike our mechanical clocks with a pendulum swinging sideways, this pendulum is swinging in an elliptical motion, the invisible rod connecting it with the sun holds our earth at its equator, and plays from tropic of cancer on one side to tropic of capricorn on the other in synchronisation with the angle of the tilt of earth's axis, marking two equal halves of the total orbital motion. The universe is full of such stellar clocks with their planetary pendulums swinging around them, as they move in space. All planets in our solar system are different pendulums attached to the sun, marking time in their respective locations.

Returning to our pendulum, its total swing is divided into four parts, taken from vernal equinox upon the equator to tropic of cancer towards the north, return to equator at autumnal equinox, from there to tropic of capricorn towards south, and finally back to the first position, moving non-stop. As the earth-swing brings the sun directly facing the vernal equinox, the perimeter of eastern hemisphere upon the earth alight with sunshine runs meeting the boundary line of arctic circle on the east while touching the antarctic circle right at its western limits effecting the phenomenon of a sunrise upon arctic and a sunset upon antarctic thresholds respectively. Thus a six-month long day begins in arctic and a night of an equal periodicity begins to set in antarctic regions simultaneously. The phenomenon repeats in an altered fashion with the returning swing reaching autumnal equinox onwards upto the tropic of capricorn, causing the antarctic witness a similar day as the arctic enters its long night. Thus a total swing both ways effects the periodicity of a daynight phenomenon in reverse fashion upon the northern and southern caps of this pendulous earth. As such this periodicity is called a solar *ahorātra*, i.e., daynight by Veda.

A solar daynight synchronises with the completion of an elliptical orbital run by the earth around the sun, a periodicity which has been named a *human year* as well. Such an orbital motion completes its run along 360°. As the said solar daynight periodicity rules only the arctic and antarctic regions the rest of our earth's 24 hourly daynights are caused by its rotational motion upon its axis, causing sunrise and sunset a daily occurrence round the year.

The orbital velocity of our earth is not uniform throughout. As the sun happens to occupy one end focus or centre of the triple-centred elliptical path, the earth's velocity keeps changing while it covers one angular degree constantly, each day. Thus its linear velocity changes, but the angular velocity remains constant and unaffected. As the earth completes over 365 daynight periods resulting from its rotational motion during one year while it completes exactly 360° along its path at its angular velocity, the Veda, on account of its invariable consistency, synchronises the periodicity of an year with the periodicity of its orbital run. The constant motion across one degree under angular velocity is called *ahorātra*, i.e., daynight, such units making a total of 360 in one year. This is generally confused with rotational daynight periods which make a total exceeding the angular number. It is, therefore, necessary that this distinction be borne in mind.

Upon the earth itself we notice that as night approaches its end, darkness in eastern sky starts getting fainter. This is an indication of the initiation of dawn which ends with sunrise. As the earth rotates, the sunlight continues moving westwards upon earth's surface with the speed of earth's rotation, constantly pursuing the tail of a dawn ahead. Vedic scholars determined that during one complete axial rotation of the earth, the dawn interval between an ending night and an oncoming day covers earth's surface longitudinally in thirty periodic leaps. Hence they divided an *ahorātra*, rotational daynight period, into thirty sub-divisions naming each a *muhūrta*. Each *muhūrta*, equalling the present 48 minutes duration, was further sub-divided into two halves, each named a *ghaṭī*. Thus 60 *ghaṭīs* or 30 *muhūrtas* equal the modern 24 hours period. A day consists of 15 *muhūrtas* and so is a night. A *muhūrta* enjoys a fairly important recognition in Vedic calendar, like an hour these days.

It may further be pointed out that our moon having axial rotation synchronized with its orbital motion periodicity around the earth which approximates 29.5 earth's daynight periods, the periodicity of sunlight's traverse upon moon's surface is also synchronised accordingly with its axial or orbital period. As such the moon's surface was also sub-divided longitudinally into 30 parts marking the traverse of sunlight upon it during its one axial or orbital period. Such a traverse section is named a *tithi*. We have a dark and a full moon on the 15th and the 30th tithi, as well as the pleasure of witnessing a growing or shrinking moon as its motion steadily hides from or exposes to earth dwellers its sunlit surface. On the contrary, the period of one degree angular motion of earth is slightly more than a *ghaṭī* larger than that of its one axial rotation. They are both called *ahorātra*, a daynight unit, or a day in common parlance, and have similar subdivisions worked out respectively described as *ghaṭī* and *muhūrta*. Such rotational subdivisions are employed in all practical purposes in respect of a rotational unit, while the academicians and astronomers use the angular subdivisions for correct calculations or effects.

The standard unit of count is an orbital round or swing of earth pendulum whether as a human year or as a solar day. Any orbital round is called a *saṁvatsara*. It has already been pointed out earlier that according to R.V. 10.190.2 *saṁvatsara* initially came to manifest from the immense ocean of radiation and particle matter that burst forth following the big bang, and is universal in its application.

Creation of a six-monthly alternate daynight phenomenon upon the earth results from a tilt in earth's axis. But even the planets having no axial tilt act as swinging pendulums along their orbital tracks in relation to their respective star suns, marking a periodicity in time which may serve as a standard with its multiples in tune with the universal time. That is why Veda declares the appearance of *saṁvatsara* alongwith the manifestation of celestial bodies, which commenced with the appearance of its hydrogen matter.

The above narration is a brief exposition of the method along which Vedic calendar has been designed. The different periodicities have been distinguished as human or solar to mark the count employed. All of these terms have been discussed and explained in chapter on kāla (time). But in this calendar a human year must not be confused with any periodicity of 365+ daynight units effected by earth's axial rotation. In this calendar a human year periodicity means one solar daynight or completion of one orbit by our earth around the sun. All subdivisions of time smaller than a *ghaṭṭī* have been left out.

VEDIC CALENDAR

Two *ghaṭīs.*	= One *muhūrta.*
15 *muhūrtas.*	= One day or night.
30 *muhūrtas.*	= One *ahorātra,* a daynight.
Sun's stay of six months north of equator.	= One *Uttarāyaṇa,* a northern solar day.
Ditto south of equator.	= One *Dakṣiṇāyana,* a southern solar day.
Completion by earth of one orbital round of the sun.	= One solar daynight or one human year.
360 solar daynights/equal number of human years.	= One *deva varṣa*, i.e., one solar year.
1200 solar years/432000 human years.	= One *Kaliyuga.*
Double *Kaliyuga*/2400 solar years.	= One *Dvāpara yuga.*
Triple *Kaliyuga*/3600 solar years.	= One *Tretā yuga.*
Quadruple *Kaliyuga*/4800 solar years.	= One *Kṛta yuga/Satya yuga.*

All four *yugas* totalling ten *kaliyugas* or 12000 solar years.	= One *caturyugī*/one *mahāyuga*.
1000 *Caturyugīs*/ $12x10^6$ solar years.	= One day of *Brahmā*.
1000 *Caturyugīs*/ $12x10^6$ solar years.	= One night of *Brahmā*.
2000 *caturyugīs*/ $24x10^6$ solar years.	= One *ahorātra*, day-night of *Brahmā*.
360 *ahorātras* of *Brahmā*.	= One year of *Brahmā*.
100 years of *Brahmā*.	= Survival period of *Brahmā*, the atom, or of the Creation.
$864x10^9$ solar years.	= Survival period of *Brahmā,* the atom, or of the Creation.
$311.04x10^{12}$ human years.	= Survival period of *Brahmā*, the atom, or of the Creation.

* The Present is the First Day of the 51st year of *Brahmā*.
* In terms of human years 1,960,853,106th human year of the said Day is running at present, which is being celebrated the world over as the international year of Physics.
* This Day is yet to run 2,359,146,894th human years by which time biological life upon earth will end.
* *Brahmā*, the 'atom' has completed **155,521,960,853,106** human years, is running in the 7th year of a new century. Its exact time can be counted upto the passing second.
* Its survivability may still continue for another **155,518,039,146,893** human years after the current one is over.
* Towards the end of great dissolution after the term of present *Brahmā*, the *Prakṛti* reconceives, so to say, effecting the recurrence of another cosmic system similar to the present one. The eternal cyclic motion of creation and dissolution continues onwards in eternity.

EPILOGUE

Veda and Modern Science, as already stated earlier, are unanimous that *Prakṛti* or energy, respective nouns used by the two, denote the substance that happens to be the fundamental cause behind the build-up of the material universe. The two are also in full agreement that this cause suffers from inertia, i.e., it is utterly incapable of acting or doing anything on its own as its nature is to resist any change in its state of being. According to modern science force and motion are the two contributory factors which play upon the energy and make it do the trick. Out of the two, force is admittedly an external agent. Regarding motion science makes no such claim. Neither it has succeeded in explaining how the inertial energy got initiated in motion. It is mum on the point.

Secondly the idle energy, even in a state of motion, with or without the help of force, remains unqualified to bring about its own transformation into an orderly, organised, smoothly working system. It possesses no intelligence for that purpose. Neither force is qualified with such a distinction.

Even an automobile needs a driver and a remote-controlled vehicle an operator. The theory of relativity and the quantum mechanics also fail to explain as their activity begins after motion has come into play. The riddle remains unsolved; neither of the two theories contribute anything to englighten the establishment of the smooth order in universe.

To highlight the paradox facing Modern Physics we again quote from the book by Stephen Hawking referred to above:

"The picture of the universe that started off very hot and cooled as it expanded is in agreement with all the observational evidence that we have today. Nevertheless, it leaves a number of important questions unanswered:

(1) Why was the early universe so hot?

(2) Why is the universe so uniform on a large scale?

Why does it look the same at all points of space and in all directions? In particular, why is the temperature of the microwave background radiation so nearly the same when we look in different directions? It is a bit like asking a number of students an exam question. If they all give exactly the same answer; you can be pretty sure they have communicated with each other. Yet in the model described above, there would not have been time since the big bang for light to get from one distant region to another, even though the regions were close together in the early universe. According to the theory of relativity, if light cannot get from one distant region to another even though the regions were close together in the early universe no other information can. So there would be no way in which different regions in the early universe could have come to have had the same temperature as each other, unless for some unexplained reason they happened to start out with the same temperature.

(3) Why did the universe start out with so nearly the critical rate of expansion that separates models that recollapse from those that go on expanding forever, so that even now, ten thousand million years later, it is still expanding at nearly the critical rate? If the rate of expansion one second after the big bang had been smaller by even one part in a hundred thousand million million, the universe would have recollapsed before it ever reached its present size.

(4) Despite the fact that the universe is so uniform and homogeneous on a large scale, it contains local irregularities, such as stars and galaxies. These are thought to have developed from small differences in the density of the early universe from one region to another. What was the origin to these density fluctuations?"

Sanctity claimed for any belief based on religion even though illogical, against public interest or against any true state of affairs in respect of any scientific fact probably became the greatest hindrance in the progress towards opening up any systematic trail towards true knowledge. In the name of God and the divine word, the Church in the West not only frowned upon but countered aggresively any knowledge discovered other than the established religious belief. And even though the spirit of Martin Luther has survived and the scientific research finally did win this battle against the Church and its influence upon the mass mind, a complex appears to have developed among the scientific community to keep their labour off God.

In India history repeated itself only with a slight yet material difference. The initiation of knowledge commencing from Veda in the dawn of history being based totally on the spirit of enquiry, as revealed by the Upaniṣadic texts, kept the scientific temper alive and kicking, giving birth to a number of schools of thought and discipline which helped the emergence of a number of *śāstras*, systematized knowledge on a subject. But the Vedic supremacy and infallibility continued as the touchstone in respect of the veracity of any *śāstra* or part thereof till about 5000 years ago. Some time after that, a communication gap occurred and even though by the time of Ācārya Cāṇakya traces of its continuance are visible, yet subsequently they appear to be lost in the mire of Vedic ignorance. In the later period the tradition of performing *yajña* degenerated into animal sacrifice and immense fiscal gains in favour of the priestly community. A revolt against the so called Vedic rituals manifested through the appearance of Buddhism and Jainism in the form of non-violence and code of conduct as preached by them. To the sufferance of Vedic cause the priestly community converted the ritualism of *yajña* performance to the worship of 24 incarnations of Viṣṇu as opposed to 24 Tīrthaṅkaras of Jain tradition and since then have been busy in trying to justify the idol worship of gods and goddesses on so called Vedic authority as some of the names do occur in Vedic texts. The priestly community even went to the extent of announcing that Veda had been lost, and granting the sanctity of scriptures to *Rāmāyaṇa, Mahābhārata, Bhāgavata* and other Purāṇas. Vedas were almost totally eclipsed and the *ārṣa* tradition, the tradition related to *ṛṣis*, was altogether substituted both in thought and performance degenerating into the form of multiple religions in the name of Veda.

Over a thousand years of political subjugation during which the power and influence of state were backing vigorous campaigns for the conversion of Indians to religions, one predominantly iconoclast while the other seeking sheep to its flock, the Indian religionists had no answer to such tirades and were busy simply preaching *kaliyuga kevala nāma adhārā*: in the present era the only refuge is reciting His name. In early nineteenth century under the British rule in India, Lord Macaulay through the instrumentation of education, set in motion the biggest campaign for the cultural conquest of India by making the coming generation forget or ridicule the past of their country and feel glorified in adopting English standards.

Then came Dayānanda challenging the worth of each and every religion, both originating in or out of India, upon the touchstone of truth. He declared that Veda is the book of all true sciences and the truth is the sole factor to determine the veracity of any religion. He had many debates called *śāstrārtha* (correct interpretation of any systematised knowledge) and scored his point. He said that all knowledge deserves to be certified to be in accordance with Veda or truth. He declared that *dharma* is the constitutional law of the universe and has been revealed to the *Maharṣis* for the benefit of humanity, codified in language in the form of Veda. According to him the Vedas as such are a proof unto themselves while any knowledge derived or earned by men from any other source whatsoever deserved to be certified by Veda. He said that a character based on education is the sole criterion of greatness or otherwise of a man. No other criterion weighs heavier than education. He maintained that all human beings are equal, but an uneducated being simply lives like an animal. When asked he declared and also wrote that he sought total political independence for India from the British rule, howsoever benevolent the latter may be, as independence far outweighs a good governance. Both Indian and foreign scholars visiting India were amazed at the depth of his knowledge and erudition without ever having had any contact with the English language and literature. Unfortunately for India and the world, his life was prematurely cut short under extremely suspicious circumstances. The mantle of his mission fell upon the Ārya Samāja, an organisation founded by him, but the same appears to have lost its track during the years of struggle for the independence of India and still appears to continue under a stupor even today.

The above narration is meant only to describe the circumstances which brought about a state of hibernation for Veda to pass for its survival. Every effort was made to destroy them both by their so called adherents as well as opponents. It was the personality of Dayānanda who brought them out of the state of hibernation described as such in a metaphorical sense. For this achievement Śrī Aurobindo of Pondicherry fame gives him full credit in following terms:

> "In the matter of Vedic interpretation I am convinced that whatever may be the final complete interpretation, Dayānanda will be honoured as the first discoverer of the right clues. Amidst the chaos and obscurity of old ignorance and age-long misunderstanding his was the eye of direct vision that pierced to the truth and fastened on that which was essential. He has found the keys of the doors that time had closed and rent asunder the seals of the imprisoned fountains."
>
> (*Bankim-Tilak-Dayānanda*, 1940 ed. page 7)

Śrī Aurobindo has also written about the claim of Maharịṣi Dayānanda regarding Veda being the book of all true knowledge and sciences. He says:

> "There is then nothing fantastic in Dayānanda's idea that Veda contains truth of science as well as truth of religion. I will even add my own conviction that Veda contains other truths of science the modern world does not at all possess, and in that case Dayānanda has rather understated than overstated the depth and range of the Vedic wisdom." (Ibid, page 67)

The purpose to mention all this is to show that such a forceful presentation of Veda by Dayānanda drew the attention of European scholars, particularly British and German who became curious to know more about them through their own studies. The politicians, particularly Macaulay, wanted them (Veda) taken apart to determine a lethal antidote for them in order to execute a cultural metamorphosis in India in the interest of British Empire. The Sanskrit-English Dictionary by Sir Monier Monier-Williams is a proof unto itself how much effort was put in to unravel the texture of the Vedic text and the literature as well as the grammar which included *Aṣṭādhyāyī, Mahābhāṣya* and *Nirukta*, the master key to Vedic study. Moreover this was done with the avowed purpose of finding their antidote for their own annihilation through tampered translation of Vedic texts as pointed out earlier. The services of Max Muller were further availed to supplement the campaign.

Whatever might have been the purpose, but such indepth study of Veda and Vedic literature put scholars wise to the wisdom contained therein at least regarding certain aspects, the study of physics being one of them. Is'nt it amazing that we find both the theory of relativity and the quantum theory make their appearance after such studies had been mulled over for decades as presented in their Vedic form. It is just not a sudden flurry of wisdom enveloping them in the 19th and 20th centuries of the Christian Era, after marking time since Aristotle for nearly 2400 years upon the two footsteps of energy and matter in the study of Physics, that we find such a brilliance enlightening the West, particularly after a two-century long loot of India in all respects. Even this sketchy presentation of Vedic Physics is enough to show that the theory of relativity is a modern version of Vedic *mahābhūta* and the quantum theory is Vedic *tanmātra* in new garb, maybe to erase Made-in-India mark.

The remarks are certainly not meant to belittle the labour put in experimentation, observation etc., and the sustained intellectual effort put in by the scientist community in the West in verifying the revelation of the leads before the same were presented publicly. All the efforts are commendable all the way. Veda is not a property belonging to India exclusively. They are a universal gift from the Supreme Being for all those who believe in It in any form, and even for those who don't. But for all the labour undertaken by the men of science as described above, such a Vedic presentation would not have come out so very well vindicated. The pity is that Vedas don't deserve to be butchered like a cow after it has been milked for some time. They are an ocean of knowledge and that credit ought to be given to them, even if for the sole reason of encouraging enterprising scholars of science to dive deep down within to collect pearls, if any, lying at the bed. Such an effort would not exhaust the Vedic store. In India vast number of people still believe Veda to be the word supreme. They (Vedas) contain nothing but the truth about the creation and everything within the creation and also about the Supreme Being. They are open to examination and test, but it would morally amount to blasphamy not to acknowledge the source of that truth. Suffice it to say that the non-acknowledgement of source has not helped the cause of science itself as a re-awakening free India would even in this dilapidated state have helped them with a lot of structural knowledge particularly in relation to Veda which would have helped prospective researchers to obtain much more milk of scientific knowledge from this most authentic ancient source.

In the light of Vedic Physics let us tackle the questions posed by Dr. Hawking in his book referred to earlier and reproduced in this chapter. But before we do that let us recapitulate the Vedic model first.

Veda declares the eternal existence of the Extraordinary Being who is Omnipresent in existence, Omniscient in knowledge and Omnificent in all creative possibilities. Such a Being is called *Saccidānanda*, the nomenclature based upon the attributes stated above. Veda calls the said Being by numerous names, each name based upon one or the other of its attributes, but they all denote the very same Extraordinary Being who according to Sāṅkhya is **sa hi sarvavit sarvakartā** (3/56), i.e., He is the all-knower and all-doer. One such name relevant for our purpose is *Puruṣa.* Veda further describes such a Being as infinite.

Veda also states *Prakṛti* as another eternal existence. But it enjoys an inertial existence on its own, although it is capable of remaining in motion once achieved if not affected by any other agent, and also to retain a material form until and unless it remains undisturbed. Under any of the three states, i.e., mere existence, motion and material formation, *Prakṛti* remains conserved, that is, it never gains or loses a bit. It is indestructible and unspendable.

According to Veda, the Extraordinary Being, which is also called *Puruṣa*, is the efficient cause of the universe while *Prakṛti* is the material cause. In other words, prompted and directed by *Puruṣa*, *Prakṛti* delivers part of itself in the form of universe.

Modern science is stuck up in determining the specifics of Creation as the same has not taken cognizance of the existence of a

Being like *Puruṣa* in Veda. A conflict between the evidence observed under theory of relativity and that under quantum theory is obstructing determination of any model. Moreover modern science is equally not clear about the end, i.e., the universe is to BE forever or it would recollapse at some point of time.

The Vedic model is a four-part model. First is the existence of *Puruṣa;* second that of *Prakṛti* whose characteristic is conservation; third is the state of their union, *Puruṣa* providing power to *Prakṛti;* and fourth is an infinite cycle of creation and dissolution in time. In mathematical terms this may be presented as follows:

1. Existence of *Puruṣa* $= \infty$
2. *Prakṛti* (in conservation) $= 10$
3. The two in juxtaposition $= 10^{\infty} = \infty$
4. Result -: (creation 1 to 9 ends in dissolution after each run for infinite count in time) $= \frac{10}{\infty} = 0$

This shows that the creation of universe would be a finite occurrence, and as the union of *Puruṣa* and *Prakṛti* is infinite the said finite occurrence of creation would continue to re-occur infinite times in succession, each time originating from and recollapsing in a state having no space-time boundary.

In the light of the above position let us now deal with Hawking questions one by one in the order they are mentioned above:

1. The Extraordinary Being *Puruṣa* which according to Veda pervades the entire *Prakṛti* and also covers the same beyond the respective and joint realms of relativity as well as quantum mechanics is certainly capable of raising the temperature from minus absolute zero to billions of Kelvin degrees simply with the application of its will. The Vedic model clarifies that present universe is not the inaugural creation. Since how long the cycle of creation and dissolution is going on, is lost in infinity? But it is certainly known to *Puruṣa* as to what ought to be the critical temperature that would neither let the radiation burst overshoot nor undershoot but to maintain its tempo last the duration of universe. The early universe was not so hot, it was just right hot under the superintendence of *Puruṣa.*

2. The big bang was not a chaotic exercise. It was duly a proper step as part of the process of the development of universe as revealed in the Sanskrit term *sphoṭa*, chapter 8. The universe looks uniform at all points of space and in all directions because, for one the bang was a properly directed excercise like a controlled experiment in a lab, and secondly *Puruṣa* was not merely present at the bang site but being omnipresent is present everywhere permeating *Prakṛti* both from within and without and fully capable of handling the dispersal of building material in all directions in the interest of keeping the process of creation on the right track. So far as there being a state of no communication among different regions was concerned as since big bang there would not have been time for light to get from one region to another, it ought to be understood from the Vedic description of *Puruṣa* that as infinite consciousness present everywhere it needed no time for knowing any state of affairs at any point. It comes to know instantaneously without taking any time whatsoever be the distance. Its existence overextends beyond the field of relativity. The whole exercise by *Prakṛti* was and is being done under the efficient direction of *Puruṣa* which pervades the space and is already present in regions being covered by galloping radiation and beyond. It needs no separate communication.
3. The third question, and all such questions which are based upon the premise of idle energy somehow getting active producing a system called universe through a million-millionth freak of coincidence simply amounts to saying that the superbly intelligent organisation apparent in the universe got manifested merely through automatic transformation of inertial, totally unintelligent, energy which for reasons unknown became activated without any activator. This would just amount to begging the question in the light of the fundamental axiom of scientific thought that something cannot be produced out of nothing. If there was no source of intelligence or wisdom or motion prior to the universe, how could such characteristics come to manifest in the system? Prof. Hawking in his quote earlier makes a proposal which conceives a state having

no space-time boundary. In pursuance of the interpretation of evidence in favour of a universe getting produced from idle energy somehow even having interactions with factors totally deprived of intelligence and wisdom, how could the universe contain such attributes and qualities? One simply fails to understand that while conceiving the possibility of a no-space-time boundary situation, how could a person of Prof. Hawking's eminence fail to visualise an omnipresent, omniscient source independent of idle energy and the rest of the contributing factors for a model of universe having such a smooth order? So far as the answer to the question is concerned the Vedic model clarifies that *Puruṣa* factor was fully capable of determining the critical rate of expansion from a zero point beyond the reach of both the quantum as well as the general relativity fields. Reference may be held of the very first *mantra* of the *Puruṣa sūkta* which specifically declares it overextending the two reaches which are ruled by the laws of creation.

4. In the infinite cycle of creations followed by dissolutions the present universe also came to be created according to a pattern or design. This one is not the same as the earlier one, but is certainly similar to its forerunner which in turn was similar to its predecessor and so on. The smoothness and the irregularities were got devised in accordance with the similarities with the earlier creations in order that a similar cosmic order may come into being. Clarifying this aspect R.V. 10/190/3 has already been discussed.

There may be other questions, no doubt. But it is humbly submitted before the scientific community that the riddle of idle energy taking to motion and a superbly orderly organised system called universe materialising through transformation do cry out the instrumentality of an omnipresent and omniscient agency quite apart and independent of the energy. If you do not feel inclined to name such an agency *Puruṣa* or *Sacchidānanda* call it x factor but assign it with all those attributes that are visible in the universe and are absent in energy in idle state. But a personal God or members of His family or staff have no scope whatsoever in Vedic Physics nor in the Science of Physics of the present day.

Before concluding, let us compare the period of the present universe as estimated by modern science and as stated by Vedic scholars. Till about the 17th century of the Christian Era the West took the earth to be not much old. But in subsequent years the scientific awakening has been responsible for increasing its age. The British scholars coming in contact with India in the 19th century were almost stupefied to learn that Indians believed the universe to be about two billion years old. By 1950 they had conceded that assessment, even though they expressed their surprise as to how the Indians in the past could think of such a long period. By the end of the 20th century their assessments had advanced to a range between 12 to 16 billion years, and around those figures they stand today, even though suggestions upto 20 billion years have been made.

Earlier in the chapter on *Kāla* the life of our universe as stated by Vedic scholars has been mentioned and the method by which the same is calculated has also been described. The assessment of 2 billion years as mentioned above was with reference to the life appearing on our earth. The life of universe is assessed far far greater than that, about 15 thousand times. This colossal figure appears to be far in excess of the latest scientific assessments or maybe staggering to many who may even be associated with science.

But in a presentation of a graphic drawing showing temperature and time in the expansion of universe, published on page 516 in Wilson's Practical Physics, the present has been marked by two arrows forward and backward and has further been specified by a dot. This dot approximates its placing around a quadrillion years. This is something quite unrelated with the text or other contents of the book. Elsewhere Dr. Jerry D. Wilson is presenting the present age of the universe to be around 16 billion years. The graphic presentation depicts a quantum jump by approximately a million times or so. There is no justification available for such a presentation in the book. Another point in this connection is that according to Vedic scholars, this universe has completed half of its life and has entered into approximately the sixth hour of the second half. Modern scientists also assess that universe has approximately covered half of its duration, even though they are not so exact in their statements as Veda is. But the difference in years on either side between the two is huge, and modern science still has a fairly long way to go, but this writer is confident that today it would hardly be called a wild conjecture even though the point fixed by Veda is thousands of times distant than its present reach.

It may also be submitted that Vedic Physics, to our mind, touches upon almost all the commanding peaks that modern science has been able to locate in this field, and presents them in proper cause and effect sequence, right from the point of its manifestation. This position leads us to only one conclusion that Vedas know what they are saying. The factors wisdom and dynamism oozing from the universe cannot be brushed aside. If afraid of the Superme Master of Sciences, use the term x factor as suggested earlier. Even a science like mathematics assumes a value to be x in order to solve a question. The solution establishes the value of initial assumption.

So far as Vedic Physics is concerned this attempt has been made to present a skeleton of the subject based upon some Vedic *sūktas* and a few stray references in *mantras* supplemented by portions from Sāṅkhya treatise and a few other references from other authentic texts. We feel we have been able to show all the structural bones of this subject as stated by modern science and the position of Veda with regard to all of them, and even parts where Vedas go beyond them. We have submitted Vedic propositions with respect not only to *Prakṛti*, the material cause, but also *Puruṣa*, the efficient cause, the *Hiraṇyagarbha*, the Vedic count of time, the *pralaya* and *mahāpralaya*, as well as the Vedic model of creative activity in mathematical terms. The touch of rancour at places is not the least against any person, race, nation or religion. Most of all it is against us, the Indians, who were the repository of Veda, the eternal fountain of knowledge, and on the basis of their wisdom were the world leader at some time in history, and later on treated the same Veda so shabbily and let others as well treat them that way, and have still, as an independent nation, kept the most superior source of scientific knowledge with them discarded on so-called religious bias. To present a few more glimpses of the treasure that the Vedas are, a few quotations from the men of learning of the modern era are being given below preceded by a brief introduction of the writer and his work.

Pandit Satyavrata Sāmaśramī was a renowned scholar of Sanskrit and English from Bengal in pre-independence India. He has written a number of books related to Veda. His famous work, *Trayī Paricaya* (*Trayī* denotes Veda and *Paricaya* is introduction) was published in Bangalā language in 1897. But the introduction of that book was in English. The following is a quote from the said introduction:

> "Our opinion is that in Vedic times our country had made extraordinary progress. In those days the sciences of Geology, Astronomy and Chemistry were called *'Adhidaivik Vidyā'* and those of Physiology, Psychology and Theology *'Adhyātma Vidyā'*. Though the works embodying the scientific knowledge of those times are entirely lost, there are sufficient indications in Vedic works of those sciences having been widely known in those days. It is needless to say that the reason why these indications are not understood now, is due to the imperfect interpretation of an expositor having no knowledge of the sciences. The study of certain portions of the Vedas leads even to the conclusion that certain scientific researches had been carried in this country to such perfection that, not to speak of this moribund country, even America, the constant source of scientific discoveries, and the advanced countries of Europe have not yet attained it. It is this which makes it impossible for us to understand the real purport of such passages. In fact, a full and satisfactory interpretation of the Veda requires a perfect familiarity with all the sciences on the part of the exposition, and it is simply a misfortune to undertake its exposition without such familiarity. What sort of exposition can one give of a book containing such words as spoon, fork, towel, etc., who does not know the use of these things? And how can a treatise containing the names of instruments used in agriculutre, which are familiar even to the childern of farmers, be understood by those rich citizens who can believe that beams can be made of paddy wood? It is perfectly plain, therefore, that it is only one that has attained a thorough knowledge of Agriculture, Commerce, Geology, Astronomy, Hydrostatics, Igneology, Botany, Zoology, Physiology and the Science of War, who can alone be a fit interpreter of the Vedas and that, it is only a commentary written

by such an expositor that can alone give full satisfaction and remove all doubts."

—Introduction to *Trayī Bhashya*- pages 8-9

Dr. V.G. Rele, a renowned scholar from Maharashtra (India) wrote a book entitled *Vedic Gods* in 1931 in which he explained at length that the gods named *Indra*, *Varuṇa*, etc., in the Veda, represent various delicate organs within the human body. In the opinion of Dr. Rele the Veda is full of the description of the sinews, glands and nerve centres of the human body and their workings. In his said work Dr. Rele, writes:

"Our present anatomical knowledge of the nervous system tallies so accurately with the literal description of the world given in the Ṛigveda that a question arises in the mind whether the Vedas are really religious books or whether they are books on anatomy and physiology of the nervous system, without a thorough knowledge of which psychological deductions and philosophical speculations cannot be correctly made. If this be true, we can surely assume that the Ancients were as far advanced in all branches of sciences as we are now; perhaps they knew much more than we know of scientific subjects and specially of the nervous system of the human body. For the true significance of some passages and the 'ṛiks' of the Ṛigveda cannot be made out because of our present imperfect knowledge of the nervous system and the difficulty still more enhanced by the symbolical aspect which the description of the anatomical facts and physiological functions wear."

—V. G. Rele in Vedic Gods

The observations of the well known historian, Śrī Pavagi of Pune, India, are worthy of note as his study regarding the Veda and the ancient civilisation in India and its history was vast and deep. He had written about two dozen books on these subjects. He had contradicted by sound arguments based upon Vedic texts the view held by Śrī B.G. Tilak that the arctic region was the land of origin of the Aryans. The following two observations are from his book written in English under the title *Vedic India Mother of Parliaments:*

"I have shown in brief how the Vedas have justly been deemed to be the real source of all knowledge and the spirit of independence."

—Pavagi's Vedic India Mother of Parliaments (1930 edition, Page 76)

"The Veda, moreover, is the fountainhead of knowledge, the prime source of inspiration, nay, the grand repository of pithy passages of divine wisdom and even eternal truth."

—Ibid. page 136

The three scholars quoted above affirm the excellence of Vedic knowledge by specifically naming over a dozen widely different sciences, from Anatomy and nervous system to Igneology and even the Science of Warfare. Mr. Pavagi affirms regarding the most important social science of Polity and Parliamentary Democracy. We have ventured to enquire in respect of Physics which has no claims to be an exhaustive enquiry, but it appears to throw some light in areas still under investigation. We strongly feel that Veda may prove a source full of promise in all respects for the humanity and human dignity and with this note we humbly end our submission.

APPENDIX: A
Glossary of Sanskrit Terms

A

adaḥ, that, a certain.

adanta, delivered, placed.

adhiṣṭhāna, substratum

ādyakāryam, (ādya initial + kāryam work) initial work performed.

agni, 1. eternally dynamic fundamental state of omnipresent state of existence, under absolute zero temperature, 2. middle mega causative out of the five of Vedic physics, denoting heat, temperature, radiation, fire, light, or any physical existence affected by such states from macro to micro levels.

agnihotra, offering oblations to fire in a ritualistic manner.

agastyaḥ, 1. the sun 2. the destroyer of all ills.

ahaṅkāra, I-ness, the sense or state of unique individual physical identity or such a feeling or state of creation.

ahorātra, (aho day + rātra night) the unit of a day and night in succession, such a periodicity in time.

ajāyata, born.

ājyam, clarified butter called 'ghṛta' in Sanskrit, oil.

akāla, (a = no + kāla = time) timeless state of existence.

ākāśa 1(ā prefix denoting the limit or boundary + kāśa becoming visible, light) the field of visibility or light to its very limits, commonly called space, or part thereof which is the site of creation, 2. the first mega causative of the Vedic classical physics, which serves as the substratum of the cosmic system, 3. also the space covered by a quantum particle of light as well as the macro field created by such quanta called photons by modern science.

ākāśa-mātram, 1. a quantum of light (photon), 2. the space (**ākāśa**) occupied by a quantum of light, i.e., a photon.

akāyam, (a = no + kāyam = body, form or figure that involves any kind of measurement) without any body, figure or form; figureless, shapeless, beyond the jurisdiction of any kind of geometry, two-dimensional, three-dimensional or multi-dimensoinal.

akṣara, (a = no + kṣara = decay) eternal, not subject to any loss or decay.

amātra, (a = no + mātrā = measure of any kind) measureless.

āmina, (also pronounced as omen) denoting supreme divinity.

aṁśa, conventionally meaning a part, this is a synonym of degree both as a part of an angle or a unit of rise and fall in temperature or pressure.

ānanda, ('a' prefix denoting whole or the limit + 'nanda' perfection in all creative activity) the limit of perfection, creativity.

anasūyā, (an = no + asūyā = grumbling in performing any duty, obligation or service) 1. conventionally meaning instant performer, 2. the name of the anti-matter particle muon.

aṅgiras, Vedic noun for neutron, denoting its birth in very high temperatures within primordial fireball.

aṅgulam, a finger, a branch.

anna, 1. a synonym of 'bramha', 2. that which grows, 3. grain.

anṛta, (an = not + ṛita = right, correct or true) that which is not ṛta, that is, not right, not correct or untrue.

antarikṣam, 1. space between the sun and the earth, 2. space between two or more celestial bodies, celestial systems, 3. literally any space in between.

apabhraṁśa, 1. fallen from any correct position, 2. in language a word deformed or corrupted.

āpaḥ, quantum activity.

apāpaviddham, (a = not + pāpa = any sin, evil deed + viddham = pierced), not pierced by any evil.

aparāmṛṣṭaḥ, (a = not + parā = thither + mṛṣṭa = affected) untouched, or unaffected in any manner whatsoever.

ardana, 1. solicitation, 2. duress.

ārṣa, by or related to a ṛṣi, seer.

aru, a fatal wound.

arundhatī, 1. (arum = a fatal consequence + dhati = inflicter), inflicter of a fatal consequence, 2. a position.

arvan, galloping radiation.

asat, (a = not + sat = existence), heterogeneous to existence, unreal.

āśaya, desire, intention .

asnāviram, (a = not + snāviram = sinews), without sinews.

asūyā, grumbling.

aśva, 1. that which pervades very fast, 2. light.

aśvognih, (asvaḥ = galloping, superfast + agniḥ = radiation) galloping radiation.

aśvinau, space between the earth and the world of light.

ātmanepadī, passive voice form of the verb.

atri, (a = not + tri = three), the sub-atomic particle muon which decays into three particles, but is taken as a single particle.

Aum, the three-letter word representing three sounds in the formation of 'Om', the ultimate Absolute.

auṇādika, 1. belonging to the group of suffixes beginining with 'uṇ', 2. a rule prescribed in that group.

āviḥ/āvis/āvir, the unmanifested state out of which something manifests

avināśī, indestructible.

āviṣkāra, occurrence of any manifestation.

avraṇam, (a = not + vraṇam = wound) one having no wound or cut.

avyaya, (a = not + vyaya = debit), undebitable, unspendable.

Āyurveda, literally meaning the science of longevity, Vedic therapeutics.

B

bāhū, arms.

bāhuvīryaḥ, one excelling in valour.

bala, force.

barhiṣi, within the inner-most heart (space).

bhāṣaṇa, denoting 'śabda', i.e. sound waves.

bhāva, abstract noun from to be, being, manifestation.

bhāvavṛttam, story of creation.

bhṛgu, plasma, fourth state of matter, a causative state of matter, extremely roasted.

bhūgola, global earth.

brahmā, atom.

brāhma, related to bramhā or atom.

brahman/brahma, literally that which grows, 1. universe 2. the undecayable, indestructible existent, 3. grain.

brāhmaṇa, a person well-versed in Vedas.

brahmāṇḍa, 1. literally the cosmic egg, 2. the cosmos.

buddhitattva, instinctive behavioural pattern in particles indicating intelligence, input ingrained in the process of creation.

C

ca, and.

catur, four.

caturmukha, having four sub-orbitals known by their first letter s,p,d,f respectively.

caturyugī, a periodicity of all the four 'yugas', eras combined, totalling 4,320,000 human years, or 12,000 solar years.

chāyā,. 1. reflection 2. mirror image of anti-matter.

cit, 1. state of wisdom and perception together with dynamism 2. quality of being fully awake both intellectually as well as for any action, 3. conscious.

D

dakṣa, deuteron.

dakṣiṇāyana, sun in the south, six-month day occurring in antarctic region.

dāna, literally donation but in the context of *yajña* this means dedicating one's own being for sublime purposes.

darśana, 1. literally seeing, but also means knowledge, 2. a treatise, 3. any one of the six treatises related to Vedas.

deśa, 1. a country or territory, 2. space.

deśakāla, territory or space + kāla = time, space-time.

deva, 1. brilliance personified, 2. a man of knowledge, 3. sunrays, 4. any object or material that reflects light, 5. any person instrumental or guide in any pursuit for knowledge in any field. 6. the sun or stars.

devānām, related or belonging to 'devas'.

devapūjā, 1. veneration generated on account of knowledge 2. honouring men of learning.

devarṣi, a term used before Vedic prajāpati Nārada known to modern science as pion (neutral) for its characteristics.

devatā, subject matter of any Vedic *mantra* or *sūkta* (Note: Two or more *mantras* may be identical in their text, but a change in respective '*devatā*' would change their object of description).

devatvam, wisdom.

devebhyaḥ, by, for and from the devas.

dharma, 1. the laws or constitution of the system of creation, 2. that which holds creation together.

dhruva, polar star.

dik, direction.

drapsam, water.

dvāpara, a period of two kaliyugas equal to 24,00 solar years or 864,000 human years.

dvyaṇuka, isotope H_2

dyāvāpṛthivī, (dyu = world of light + pṛthivī = earth) universe as a whole.

dyu, the world of light.

dyusthānīya, sunlocated, starlocated.

dyuti, luminescence

G

gamana, motion.

gandha, 1. the attribute of the fifth mega causative *pṛthvī* denoting solicitation by way of seeking to form a bond, 2. holding mobile atoms or particles thereof in a bond.

gandhana, denoting four kinds of forces, namely, gravitational, electromagnetic, strong nuclear and weak nuclear, covered under the common Vedic noun Vāyu.

gandharvaveda, the upa-veda of Sāma Veda, an authority on music.

garbha, a womb or interior.

gati, 1. literally motion, 2. in Sanskrit grammar the roots so specified denote knowledge, motion, accomplishment, both collectively and severally.

gatikaraṇa, creation of motion or motions.

gāvaḥ, 1. cows, 2. mammals, 3. entities in motion.

ghāta/ghātana, with the addition of prefix *sam* and *vi* respectively denotes fusion and fission, the inter-action of 4th mega causative Jala.

ghaṭī, a periodicity of time equivalent to modern 24 minutes.

goda, guardian or protector of universe.

grahaṇa, to take or hold.

grīṣma, summer.

guṇa, attribute.

guru, teacher.

H

ha, definitely.

haviḥ/havir, 1. exchange of particle pi-meson back and forth under strong nuclear force, 2. oblations.

havirbhū, anti-matter pion.

hiṁsā, violence, duress.

hiṁsana, strong nuclear force.

hiṁsana-mātram, pi-meson as quantum particle for strong nuclear force.

hiraṇya, light.

hiraṇyagarbhaḥ, literally meaning the fetus of light, (there are four mantras, totally identical, in this respect, but each carries a different devatā, i.e., its subject matter).

hiraṇyamayeṇa, made of light.

hiraṇyaśṛṅgaḥ, literally meaning the peak of light, the supernova state.

hotā, a performer of *yajña*.

I

idam, this.

Indra, electricity.

Īśvara, the Lord.

J

jaḍa, affected by inertia, inertial.

jajñire, appeared from the *yajña*.

jala, fourth *mahābhūta* or mega causative which effects fusion/fission.

jala- mātram, quantum state of jala.

janma, a synonym of water, literally birth.

jīva-pralaya, a state of part dissolution denoting disappearance of the biological creation.

jñāna, knowledge.

K

ka, 1. name of the metaphorical zygote formed and developed within the hiraṇyagarbha, the quantum state of the universe, 2. light inclusive of heat and splendour, 3. prajāpati, brahman, growth, agni, and time, etc.

kāla, time.

kali, literally meaning one; it denotes the basic unit of time of 1200 solar or/ 432,000 human years.

kaliyuga, (kali = one + yuga = a periodicity of 1200 solar years / 432,000 human years.

kalpa, a periodicity of the day or night of brahmā,. i.e., atom, equal to ten thousands kaliyugas.

kardama, anti-matter.

karaṇa, doing, to do.

kāraṇa, a cause, such as *upādāna kāraṇa*, i.e., material cause.

karma, 1. action, 2. performance.

kaṭhopaniṣad, the upaniṣad by this name.

katidhā, how, in what way.

kaviḥ, omniscient.

kāvya, poetry, work par excellence.

kenopaniṣad, a leading upaniṣad.

khagola, the globe of ākāśa, i.e., space.

khaṇḍa-pralaya, part dissolution, partial annihilation of the universe.

kim, interrogative what.

kleśa, affliction, distress, anguish.

kratu, proton.

kṛta, 1. a periodicity equal to four kaliyugas/ 4800 solar years/ 1728000 human years.

kṛti, an act of doing, making, performing, composing, 2. a product.

kriyā, anti-matter particle of kratu i.e., anti-matter proton.

kṣetra, a field.

kṣobha, a disturbance, an agitation.

L

lakṣmī, prosperity, material wealth.

laya, molten state, dissolution.

M

madhyamātram, joule, unit of electrical charge.

madhya-sthānīya, 1. mid-based, 2. based in the space between the sun and the planet earth, 3. based in inter-space anywhere (for electricity, i.e., 'agni').

mahābhārata, the famous Indian Epic.

mahābhūta, mega causative of Vedic physics.

mahābhūta agni, the third mega causative of matter.

mahābhūta ākāśa, the first mega causative of matter.

mahābhūta jala, the fourth mega causative of matter.

mahābhūta pṛthvī, the fifth mega causative of matter.

mahābhūta vāyu, the second mega causative of matter.

mahadākhyam, (mahat + ākhyam), called or named mahat.

mahādeva, the great illuminated one, ākāśa (space).

mahaḥ loka, inner galactic space of our milky way galaxy.

mahāpralaya, the great dissolution, the re-collapse, the big crunch.

Maharṣi Vedavyāsa, the famous author of epic Mahābhārata, and a great scholar of Vedas, better known by his title Vedavyāsa.

mahat, 1. initial work done by *Prakṛti* (energy), 2. the primordial fireball

mahat sphoṭa, 'big bang' in modern scientific terminology, but the blast for developing into enormous proportions according to Vedic Physics.

mahimā, greatness.

maṇḍala, a section of the R.V. text containing a number of *sūktas*.

manīṣī, highly learned being.

mantra, an instrument of thought, in verse or prose, belonging to Veda.

mantradraṣṭā, the seer of the knowledge encapsuled within any *mantra*.

manvantara, a long periodicity of time.

marīci, a particle of light, a photon.

mārtaṇḍa, 1. born of a dead star, 2. our sun.

marut, a sub-atomic particle that decays.

marya, a subatomic particle belonging to a different group, which also decays.

maryādā, limit, boundary.

mātrā, quantity, measure.

mātram, the quantum state.

māyā, capable of assuming form.

mitthyā, false, illusion.

mitra, hydrogen.

mitrāvaruṇau, hydrogen and oxygen.

muhūrta, a periodicity of time equal to 48 minutes

mukha, a sub orbital

N

naimittika, 1. instrumental for the purpose, 2. related to the Puruṣa in respect of Creation.

nārada, subatomic particle pion (neutral).

nāśa, 1. destruction of form, 2. disappearance.

neti, (na = no + iti = end) no end.

Nighaṇṭu, the initial part of Nirukta containing groups of synonymous Vedic terms which have been subsequently explained.

nirguṇa, having zero attribute.

Nirukta, literally meaning 'explicitly explained'. This is a branch of the Vedic study in which only one work authored by Yāska is now available.

O

Om, 1. name of the Supreme Master of Sciences, 2. the Ultimate Absolute.

oṣadhīḥ, atomic fuel.

P

pāda, a step.

padārtha-vidyā, science of physics.

pala, a small unit of time.

pañca, five.

Pāṇini, the famous Grammarian of Sanskrit language.

parama, absolute.

parama gurutva, 1. parama = absolute + gurutva = gravity, 2. the zenith of gravity.

paramāṇu, 1. the smallest particle of matter, 2. a unit of time in which a ray passes a neutrino.

parama śūnya, absolute zero.

Paramātmā, Absolute Being in eternal motion.

Parameśvara, parama = absolute + Īśvara = Master or Lord.

paribhūḥ, is/ being all around.

pārśve, two sides.

paśum, 1. that which is tied within one's self by being meditated upon by wise persons, 2. The Saccidānanda, 3. The Puruṣa.

patnī, 1. literally a wife, but truly the female participant in the performance of a *yajña*, 2. a companion in a *yajña*, 3. a companion in the performance, woman of the cosmic *yajña*.

Pīlu, a molecule.

pra, a prefix denoting 'onwards'.

prajāpati, 1. procreator sub-atomic particle, 2. initial state of procreation.

prakāśana, 1. manifestation through weak force.

prakāśanamātram, particle, quantum state of the weak force.

prakṛti, energy in modern physics.

pralaya, pra = onwards + laya = dissolution.

pra-mātram, abbreviation of 'prakāśana mātram'.

prāpti, 1. attainment, 2. the third realm of meaning denoted by any Sanskrit root specified to mean gati.

prathana, 1. expansion 2. the process of molecule formation.

pṛthivī/pṛthvī, the fifth mega causative effecting molecule formation.

pṛthvīmātram, electron as quantum particle in molecule formation.

pṛthvī-sthāniya, earth-based.

pūjā, act of reverence, honour.

pulaha, sub-atomic particle neutrino.

pulasti, sub-atomic particle pion.

pūrṇa, self-contained whole, full

purohita, one who leads or is always ahead in bringing welfare to all.

Puruṣa, the Instrumental or Efficient Cause of Creation.

puruṣa, a human being.

Puruṣa sūkta, 1. a *sūkta* having a group of *mantras* describing *Puruṣa,* in each Veda, 2. chapter 31 of Y.V.

Puruṣa Viśeṣaḥ, Extraordinary Being / The Supreme Master.

pūrve, earlier (in the *mantra*).

puṣkara, atmosphere.

pūtadakṣam, 1. expert in cleansing, 2. cleansing agent.

R

rājanya ḥ, men of valour, persons who save the weak and oppressed from the mighty oppressors.

rajas, the second attribute of *Prakṛti,* energy, which denotes motion.

rājasūya yajña, a specific *yajña* provided in Veda to be performed by kings and emperors.

rākṣasāḥ, 1. the demons, 2. parasites upon human beings, 3. reverse of sākṣarāḥ.

rasa, the attribute of the fourth mega causative, *jala,* culminating the effect caused.

rāṣṭra, a nation, a politically well knit society with distinct characteristics denoting its ideals and way of living.

raśmiḥ, 1. a ray of light, 2. a measuring string, 3. a rein.

ratnadhātamam, supreme capable to create jewels.

rayi, wealth, riches, property dedicated for benefic deeds and social good.

Ṛgveda, the first and the biggest Veda containing *mantras* stating the attributes, action/re-action and specific nature of the creative ability.

ṛṣi, a seer of the knowledge contained in Vedic *mantras.*

Ṛṣi Viśvāmitra, a very famous *ṛṣi* believed to be the first to place the first stationary satellite around the earth.

ṛta, 1. *prakṛti* in motion, 2. kinetic energy, 3. the creative order in the universe, 4. laws of creation or any creative order.

ṛtvij, a priest at the performance of a *yajña.*

rūpa, 1. visible image, 2. attribute of 3rd mega causative.

rūpakriyā, the process of creating image.

S

sa, He, That Supreme Being.

śabda, 1. attribute of first mega causative *ākāsá,* 2. electromagnetic waves and sound waves.

Saccidānanda, 1. sat=existence + cit = wisdom/dynamism/consciousness + ānanda = total creativity, so named on the basis of these attributes, 2. Supreme Master of Sciences, 3. the Efficient Cause.

sādhyāḥ, men of wisdom.

sahasasputra, 1. belonging to vāyu (force) + putra = like an offspring, 2. existing like an offspring of 'vāyu' (force), i.e., 'agniḥ', the 3rd mega causative.

sahasra, 1. the universe, 2. innumerable.

sākṣarāḥ, merely literates.

sākṣī, a witness.

sama, exactly the same, equal.

sāma, *mantras* from the Sāmaveda, witch are prayers.

samādhi, 1. state of deep meditation, 2. state of stabilised mental vision, 3. state of extreme concentration, in which the psyche becomes oblivious of all save only the one it wants to know or see.

samanta, all around.

samanta maryādā, limit all around.

Sāmaveda, one of the Vedas, having *sāmas, mantras* of prayer.

sambhūti, anti-matter photon.

samidhaḥ, 1. specified wood that easily gets consumed by getting ignited in a *yajña*, 2. twenty-one items mentioned in the conduct of cosmic *yajña.*

samṛddhiḥ, prosperity and abundance.

saṁsleṣaṇa, mixing together as one stuff.

saṁvatsara, 1. earth's one complete orbit around the sun, counted as one human year, 2. a unit of time and distance based on any such event.

sāmyāvasthā, state of conservation.

saṅgati, unified motion.

saṅgatibhāva, (saṅgati = unification of motions+bhāva = occurrence) occurrence of unification of motions.

saṅgatikaraṇa, unification of motions.

sāṅkhya, a Vedic treatise which gives the count of the causative elements of the cosmic system.

saṅkhyāna, counting.

saptaraśmiḥ, (sapta = seven + raśmiḥ = a ray, a measuring cord), 1. seven colour braided ray, 2. seven colour braided measuring cord of time.

saptarṣi, seven sub-atomic particles.

śarad, autumn.

sarvahut, all-consuming cosmic *yajña,* process of creation.

śāstra, 1. any systematised knowledge, 2. any book containing such knowledge.

śāstrārtha, any debate to determine what is in consonance with śāstras.

sat, 1. present, 2. eternally present, 3. in eternal existence.

śatapatha, each Veda has a Brahmaṇa. Śatapatha is a Brāhmaṇa of Yajurveda.

sattva, first attribute of prakṛti denoting existence.

satya, matter.

Sāyaṇa, a learned person who has written commentaries on Vedas.

śiva, the ākāśa stretching all around containing the universe.

soma, the cooling stabilizing effect of 4th *mahābhūta jala.*

sparśa, touch attribute of 2nd mega causative *vāyu,* denoting its quantum activity effect through touch.

sphoṭta, 1. development or growth in a tremendous burst, 2. Vedic 'big bang'.

śrat, that which survived intense burning and roasting.

śraddhā, anti-matter neutron.

śravaṇa, the act of hearing.

śrī, splendour.

sṛṣti, creation.

sṛṣti vidyā, the science of physics.

sthūla, mega, increased in size.

sthūlabhūta, mega causatives.

stuta, stated.

stuti, statement.

sūcana, 1. gravitation, 2. keeping informed.

sūcanamātram, graviton.

śuddham, pure, faultless. accurate.

śūdra, non-specialised being visualised as the pāda (feet), specified by their capability for providing motion to the Raṣṭra Puruṣa, work force.

śukram, the omnipresence of heat below absolute zero temperature.

sūkta, literally 'su' = well + ukta = said, a group of Vedic *mantras.*

śūnya, zero

sūra, the sun or the stellar system.

sūtra, an aphorism.

svabhāva, exclusive individuality.

svayambhū, birthless.

T

tadānīm, at that time.

tamas, condensation or fusion of K.E. into matter, 2. third attribute of *prakṛti.*

tanmātra, quantum state of each of the five mega causatives.

tāpa, heat.

tāpamātram, calory, quantum state of heat.

tapas, heat.

tasmāt, from It, from the Puruṣa.

tejas, brilliance.

tithi, periodicity of sun's light traversing 1/30 of lunar surface longitudinally.

tryaṇuka, a tritium, H_3.

tretā, a period equal to three times that of a 'kaliyuga', i.e., 1200 x 3 solar years or 432000 x 3 = 1296000 human years.

tri, three.

triṇābhi, triple-centred, an ellipse.

triṇabhi-cakram, for purposes of calculation an ellipse may be treated equivalent to a circle imagined drawn from the middle centre with a mean of two other radii.

tri-śaṅku, three-stage conicular rocket vehicle.

tuccha, a zero.

Tvaṣṭā, literally a carpenter, but here a synonym for Viśvakarmā/ Puruṣa.

U

udaka, water.

ūhā, visionary conjecture.

upādāna kāraṇa, material cause.

upaniṣad, many writings that reveal the mystery that lies or rests underneath the eternal system of things.

ūru, shanks, the parts which support upper structure.

urvaśi, electricity, lightning.

ūṣmā, temperature.

ūṣmamātram, quantum state of temperature.

utsāha, electromagnetic force in attraction and repulsion.

utsāhamātram, quantum state of electromagnetic force.

uttarāyaṇa, sun's six-month sojourn in northern hemisphere.

V

Vaiśeṣika, one of Vedic treatises, other than Sāṅkhya.

vaiśya, one who is devoted to earning wealth by trading, described to form the shanks as part of the personification of the Puruṣa.

varṇa, psychological shade of personal nature.

varṇa-vyavasthā, varṇa = psychological nature+vyavasthā = system, a social system prescribed by Veda based on four divisions on the basis of human nature.

varuṇa, 1. oxygen, 2. oxidising agent.

vasanta, spring season.

vasiṣṭha, 1. water as a superior support resource for habitation, 2. electron.

vasu, habitable planets.

vāyu, single Vedic noun for all of the four kinds of forces, namely gravitational, electromagnetic, strong nuclear and weak nuclear force of modern physics.

vāyumātram, quantum state of force.

Vedavyāsa, one who was highly proficient in Vedic literature and texts.

Vedārthadīpakaḥ, a title given to *Nirukta,* denoting the same to be the lamp that enlightens the meanings contained in Veda.

Vedic koṣa, a modern dictionary of Vedic terms.

vimāna, an aircraft.

vināśa, destruction.

vipāka, effect, result, consequences.

vipala, a minute periodicity of time.

viṣṇu, 1. omnipresent, 2. yajña.

visphoṭa, a bang, an explosion.

viśva, all, total creation, universe, cosmos.

viśvakarmā, creator of the universe.

vraṇa, a wound.

vyoma, space, ākāśa.

Y

yācanā, solicitation.

yajña, a noun capable of expressing the central thrust of entire Vedic knowledge and thought. It has been briefly explained in a chapter.

Yajña-Puruṣa, 1. the efficient cause of the universe, 2. the universe itself.

Yajurveda, the second Veda related to Yajña.

yakṣa, spiritual apparition.

yāñcā, to solicit, solicitation.

yantra-sarvasva, title of an ancient manuscript, now lost, meaning total mechanics.

yaugika, etymological.

yoga, not the conventionally known set of physical exercises, but the highest practical discipline of mind and spirit advocated by Veda.

yogarūḍha, a specific meaning covered under the range of etymological meanings.

yuga, a mundane period of 1200 solar or 432,000 human years as the first unit, its double, triple and quadruples.

yuti, a conjunction of planets of our solar system.

APPENDIX: B
Bibliography

1.	Ṛgveda	Respective Texts, first edition, published by Harayānā Sāhitya Prakāśana, Rohtak (India)
2.	Yajurveda	
3.	Sāmaveda	
4.	Atharvaveda	
5.	Sāṅkhya Darśana (Kapila),	Commentary by Āryamuni, published by Harayānā Sāhitya Sansthāna, Rohtak in 1976.
6.	Vaiśeṣika Darśana (Kaṇāda)	—do—
7.	Nirukta commentary by Chandramaṇi Pāliwāla,	First edition, published by Ārya Kanyā Gurukula, Narelā, Delhi.
8.	Aṣṭādhyāyī (Pāṇini)	
9.	Yajurveda Bhāṣya (Dayānanda)	Published by Paropakāriṇi Sabhā, Kesarganj, Ajmer.
10.	Ṛgvedādibhāṣya Bhūmikā (Dayānanda)	—do—
11.	Satyārthaprakāśa (Dayānanda)	Published by Ārsha Sāhitya Prakāśan Trust, Khāri Bāoli, Delhi, Aug. 1975.
12.	Satyārthabhāskara (Vidyāṇanda)	First edition, published by International Aryan foundation, 302, Captainvilla, Montmary Road, Bandra, Mumbai-400050
13.	Viśvakoṣa,	Famous Sanskrit Dictionary.
14.	Vedic kosha (Rajvira Sastri),	First edition, published by Ārsha Sāhitya Prakāśan Trust, Khāri Bāoli, Delhi.

15.	Amara Kosha (Amar Singh),	First edition, published by Chaukhamba Sanskrit Series office, Chauk, Vārāṇasi
16.	Sanskrit-English Dictionary (Sir M.M. Williams)	Oxford II edition republished by Motilal Banarsidass, 41, UA Bungalow Road, Delhi-110007.
17.	English-Sanskrit Dictionary (Sir M.M. Williams)	—do—
18.	The Students Sanskrit-English Dictionary (V.S. Apte),	Published by Motilal Banarsidass as above.
19.	The Random House Dictionary of the English Language, College edition	Reprint by Allied Publishers Pvt. Ltd., 1975, New Delhi.
20.	Webster's New Collegiate Dictionary,	Indian Edition, 1973. Scientific Book Agency, 22, Raja Woodmunt Street, Kolkata (India)
21.	Īśāvāsyopanishad	Gīta Press, Gorkhpur
22.	Taittirīyopanishad	—do—
23.	Kenopanishad	—do—
24.	Kaṭhopanishad	—do—
25.	Śvetāśvataropanishad	—do—
26.	Śatapatha Brāmhaṇa	—do—
27.	Taittirīya Brāmhaṇa	—do—
28.	Bṛihadāraṇyakopaniṣad	Vaidic Pustakalya, Ajmer
29.	Practical Physics (Jerry D. Wilson),	CBS College publishing, Philadelphia/New York.
30.	A Brief History of Time (Stephen Hawking)	
31.	'Q' is for Quantum A to Z particle Physics (John Gribbin),	Universities Press (India) Hyderabad.
32.	Dhātupāṭha (Pāṇini),	Elaborated in 'Ākhyātika' (Dayānand) published by Paropkāriṇī Sabhā, Kesarganj, Ajmer (India).

33. Mahābhārata (S.D. Satwalekar), Published by Swādhyāya Mandal, Pārdi, Distt.- Balsād (India) in 1972.

34. Śrīmadbhagvadgītā, Published by Gīta Press, Gorakhpur.

35. Sabdakalpadrumaḥ, Nāg Publishers, Delhi-7.

36. Vedic Kosa (Upādhyāya) —do—

37. Bārhaspatyam, By Chaukhamba Sanskrit series, Vārāṇasī (India)

38. Śrīmad Bhāgavata Mahāpurāṇa, Gīta Press, Gorakhpur (India)

39. Vedic Sampatti, Published by Sārvadeshika Ārya Pratinidhi Sabhā, Dayānand Bhawan, Ansari Road, New Delhi.

40. Bṛihad Vimāna śāstra, —do—

41. Bible in India, As reported by Bhawani Lal Bhartiya, Jodhpur (India)

42. A few issues of 'Scientific American' Published from New York.

43. The paper read by Dr. Mahavira New Delhi.

SUBJECT INDEX

I

J

K

L

M

N

O

P

R

S